AF745109

Number Theory - Classical Foundations and Modern Perspectives

Edited by Bruno Carpentieri and Mudassir Shams

Published in London, United Kingdom

Number Theory - Classical Foundations and Modern Perspectives
http://dx.doi.org/10.5772/intechopen.1009697
Edited by Bruno Carpentieri and Mudassir Shams

Contributors
Altaf Ahmad Bhat, Audace A. V. Dossou-Olory, Bruno Carpentieri, Fairouz Tchier, Faisal Sultan, Ilaria Viglino, Javaid Gani Dar, Kelvin Tafadzwa Mpofu, Mircea Ghidarcea, Mohammad Younus Bhat, Mudassir Shams, Muhammad Usman Ghani, Patience Mthunzi-Kufa, Sodiq A. Mojeed, Syed Ajaz K. Kirmani, Tim Rowland

© The Editor(s) and the Author(s) 2026

The rights of the editor(s) and the author(s) have been asserted in accordance with the Copyright, Designs and Patents Act 1988. All rights to the book as a whole are reserved by INTECHOPEN LIMITED. The book as a whole (compilation) cannot be reproduced, distributed or used for commercial or non-commercial purposes without INTECHOPEN LIMITED's written permission. Enquiries concerning the use of the book should be directed to INTECHOPEN LIMITED rights and permissions department (permissions@intechopen.com).

Violations are liable to prosecution under the governing Copyright Law.

Individual chapters of this publication are distributed under the terms of the Creative Commons Attribution 4.0 License which permits commercial use, distribution and reproduction of the individual chapters, provided the original author(s) and source publication are appropriately acknowledged. If so indicated, certain images may not be included under the Creative Commons license. In such cases users will need to obtain permission from the license holder to reproduce the material. More details and guidelines concerning content reuse and adaptation can be found at http://www.intechopen.com/copyright-policy.html.

Notice

Statements and opinions expressed in the chapters are these of the individual contributors and not necessarily those of the editors or publisher. No responsibility is accepted for the accuracy of information contained in the published chapters. The publisher assumes no responsibility for any damage or injury to persons or property arising out of the use of any materials, instructions, methods or ideas contained in the book.

First published in London, United Kingdom, 2026 by IntechOpen
IntechOpen is the global imprint of INTECHOPEN LIMITED, registered in England and Wales, registration number: 11086078, 167-169 Great Portland Street, London, W1W 5PF, United Kingdom

For EU product safety concerns: IN TECH d.o.o., Prolaz Marije Krucifikse Kozulić 3, 51000 Rijeka, Croatia, info@intechopen.com or visit our website at intechopen.com.

British Library Cataloguing-in-Publication Data
A catalogue record for this book is available from the British Library

Number Theory - Classical Foundations and Modern Perspectives
Edited by Bruno Carpentieri and Mudassir Shams
p. cm.
Print ISBN 978-1-83635-626-4
Online ISBN 978-1-83635-625-7
eBook (PDF) ISBN 978-1-83635-627-1

If disposing of this product, please recycle the paper responsibly.

IntechOpen
intechopen.com

Built by scientists, for scientists

Explore all IntechOpen books

Meet the editors

Bruno Carpentieri received his Laurea in Applied Mathematics from the University of Bari, Italy, and a Ph.D. in Computer Science from INPT, France. He held postdoctoral positions in Austria and Italy before joining academia. He has served as an Assistant Professor at the University of Groningen and as a Reader at Nottingham Trent University, and since 2017 has been an Associate Professor of Applied Mathematics at the Free University of Bozen-Bolzano. His research focuses on numerical linear algebra, high-performance computing, and computational mathematics, with applications in scientific modeling and engineering. He has served on scientific advisory boards, editorial committees, and review panels for leading journals and conferences, supervised over 20 student projects at various levels, and authored approximately 80 peer-reviewed publications and two books.

Mudassir Shams received his Ph.D. in Computational Mathematics (Numerical Analysis) from Riphah International University, Islamabad, Pakistan. He is currently serving as an Assistant Professor at the Department of Mathematics, Faculty of Arts and Science, Balıkesir University, Balıkesir, Turkey. Before joining Balıkesir University, he worked as a Research Assistant at the Faculty of Engineering, Free University of Bozen-Bolzano, Italy, and served as an Assistant Professor at Riphah International University, Islamabad, Pakistan. His research interests include computational mathematics, particularly numerical analysis, computational fluid dynamics, iterative methods for nonlinear algebraic and transcendental equations, integral inequalities, and numerical techniques for solving initial and boundary value problems in both classical and fractional frameworks. Dr. Shams has 82 publications, including a book, book chapters, and numerous journal articles in high-impact journals. He has supervised one Ph.D., more than 28 MPhil, and 30 BS Mathematics students. He serves as a reviewer for various international journals and has presented papers at several international conferences as an invited speaker. His significant contributions have advanced the field of computational mathematics and its applications.

Contents

Preface

Number theory is a fundamental area of mathematics with a long history and a wide range of connections to other disciplines. Originating in the study of integers, divisibility, and prime numbers, it has developed into a field that interacts with areas such as algebra, combinatorics, cryptography, and theoretical computer science. In recent years, computational and algorithmic approaches have further expanded their scope, contributing to both theoretical investigations and practical applications.

The present edited volume, **Number Theory – Classical Foundations and Modern Perspectives**, brings together a collection of contributions that reflect different aspects of contemporary research and education in number theory. The chapters address classical theoretical questions, modern computational methods, interdisciplinary applications, and pedagogical issues. The volume is intended to provide a representative overview of current themes and approaches rather than a unified treatment of a single topic.

The book is organized into three sections.

The first section, **Classical and Structural Foundations in Number Theory**, focuses on foundational topics and structural aspects of the subject. The chapter *An Average Version of the Chebotarev Density Theorem for Families of Galois Extensions* studies families of number fields arising as splitting fields of polynomials and establishes an average form of the Chebotarev Density Theorem, including probabilistic results on the distribution of prime splitting types. The chapter *Perspective Chapter: Experimental Insights on Prime Gaps* presents computational investigations of prime gaps based on large-scale sieving techniques, highlighting observed statistical patterns and related conjectural questions. The section concludes with *Matula Numbers and their Applications*, which examines a bijection between rooted trees and positive integers and discusses its applications in combinatorics and chemistry.

The second section, **Computational, Applied, and Algorithmic Perspectives**, addresses number-theoretic methods and ideas in computational and applied settings. The chapter *On Efficient Single-Step Scheme for Solving* ***p****-Adic Polynomial in* $\mathbb{Z}_p$*: Convergence and Fractal Analysis* introduces an iterative method for solving nonlinear equations in the ring of p-adic integers, with an analysis of convergence properties and associated dynamical behavior. The chapter *Post-Quantum Cryptography: Number Theoretic Foundations and Future-Proof Protocols* surveys cryptographic schemes designed to resist quantum attacks, outlining the number-theoretic assumptions underlying lattice-based, code-based, multivariate, and isogeny-based approaches. The chapter *Ghani Mersenne and Fermat Entropy for Some Dendrimers and Their Structure-Property Modeling* proposes numerical descriptors inspired by Mersenne and Fermat numbers and applies them to the analysis of dendrimer structures. The section concludes with *Slepian Pollak Inequality for Multi-Band Wavelets*, which studies an energy inequality for multi-band wavelets and discusses its relevance in signal analysis.

The third section, **Perspectives on Proof, Education, and Pedagogy**, is devoted to reflective and educational aspects of number theory. The chapter *Perspective Chapter: Proofs in Number Theory – Audiences, Purposes, and Modes* examines different modes of proof through examples drawn from number theory, with particular reference to established frameworks in mathematics education.

Although the emphasis of the volume is on research-oriented contributions, several chapters also address educational considerations, including the communication of mathematical ideas and the use of computational tools in teaching. The book is intended for researchers, graduate students, and educators interested in number theory and its various applications.

The editors wish to thank all contributing authors for their cooperation throughout the preparation of this volume. The editors also acknowledge the continuous professional support of Ms. Mia Vulovic, Publishing Process Manager at IntechOpen, for her assistance and guidance during the editorial process.

Bruno Carpentieri
Faculty of Engineering,
Free University of Bozen-Bolzano,
Bolzano, Italy

Mudassir Shams
Faculty of Arts and Science,
Department of Mathematics,
Balıkesir University,
Balıkesir, Turkey

Section 1

Classical and Structural Foundations in Number Theory

Chapter 1

An Average Version of the Chebotarev Density Theorem for Families of Galois Extensions

Ilaria Viglino

Abstract

We discuss the arithmetic of certain families of number fields, obtained as the splitting fields of families of polynomials. The first and main example is the family of monic, integer polynomials over a fixed number field extension of fixed degree, with bounded height. It was van der Waerden who first proved that almost all (in a suitable sense) such polynomials have splitting field over the rationals with Galois group isomorphic to the symmetric group. On the other hand, all these kinds of number field extensions arise in this way for some polynomial as above. We show an average version of the Chebotarev Density Theorem for this family of polynomials. In particular, this gives a Central Limit Theorem for the number of primes with given splitting type in some ranges.

Keywords: analytic number theory, Chebotarev density theorem, class number, large sieve, symmetric group

1. Introduction

The Chebotarev Density Theorem stands as one of the cornerstones of algebraic number theory. It governs how prime ideals of a number field are distributed among the conjugacy classes of its Galois group, providing a unifying generalization of classical splitting results for primes in the extensions of $\mathbb{Q}$. Through this correspondence between arithmetic and group structure, the theorem describes the frequency with which Frobenius elements fall into specific conjugacy classes, thus encoding deep information about the factorization of primes in number fields.

Beyond individual instances, much recent progress has focused on *average* versions of Chebotarev-type statements. Studying the expectation and variance of splitting behaviors across large families of polynomials or fields reveals how typical Galois and splitting phenomena manifest statistically. This shift from individual to collective behavior brings into focus universal patterns that are obscured when treating one field at a time. Such average results form a major theme of modern arithmetic statistics, blending analytic number theory, probability, and algebraic geometry to describe how algebraic data are distributed in large families.

A natural framework for these investigations comes from families of monic integral polynomials. Each polynomial defines an extension of $\mathbb{Q}$ (or of a general base field K), and the distribution of its coefficients provides a natural random model.

The factorization of a polynomial modulo primes then reflects the cycle structure of Frobenius elements in its splitting field, creating a direct bridge between Galois theory and residue arithmetic. This perspective extends classical ideas—such as van der Waerden's study of reducible equations over the integers—into a modern statistical setting where one seeks asymptotic densities and probabilistic laws.

Recent work has refined these probabilistic models by combining large-sieve estimates, geometric methods, and explicit bounds on error terms. These advances have led to sharp quantitative statements on how often certain Galois groups occur and how primes split on average within random families of fields. Such results not only deepen our theoretical understanding but also have computational consequences: They inform algorithms for constructing number fields, factoring polynomials, and designing cryptographic primitives whose security relies on number-theoretic randomness.

1.1 Counting S_n-polynomials

Let $K/\mathbb{Q}$ be a number field of degree d. For integers $n \geq 2$ and $N > 0$, consider the set

$$\mathscr{P}_{n,N}(K) = \{f(X) = X^n + \alpha_{n-1}X^{n-1} + \cdots + \alpha_0 \in \mathcal{O}_K[X] : \mathrm{ht}(f) \leq N\},$$

where $\mathrm{ht}(f)$ denotes the maximum absolute value (in a fixed integral basis) of the coefficients of f. Writing each coefficient as

$$\alpha_k = \sum_{i=1}^{d} a_i^{(k)} \omega_i, \qquad a_i^{(k)} \in \mathbb{Z},$$

for a chosen $\mathbb{Z}$-basis $(\omega_1, \ldots, \omega_d)$ of $\mathcal{O}_K$, we view the integer parameters $a_i^{(k)}$ as independent random variables uniformly distributed in $\{-N, \ldots, N\}$. This probabilistic interpretation allows us to discuss asymptotic densities within $\mathscr{P}_{n,N}(K)$ as $N \to \infty$.

Denote by $\mathscr{P}^0_{n,N}(K)$ the subset of $\mathscr{P}_{n,N}(K)$ consisting of those f whose Galois group over K is isomorphic to S_n. Such polynomials will be called *S_n-polynomials over K*. Classical results show that almost all polynomials of fixed degree have full symmetric Galois group. Quantitatively, one has

$$\frac{|\mathscr{P}^0_{n,N}(K)|}{|\mathscr{P}_{n,N}(K)|} \to 1 \qquad \text{as } N \to \infty.$$

For $K = \mathbb{Q}$, this principle goes back to van der Waerden [1], who obtained an explicit error term of size $O(N^{-1/[6(n-2)\log\log N]})$. Gallagher [2] later improved this estimate *via* the large sieve to $O(N^{-1/2}\log N)$, and Dietmann [3] refined it further using resolvent techniques to $O\left(N^{-2+\sqrt{2}+\varepsilon}\right)$ for any $\varepsilon > 0$. A conjecture of van der Waerden, asserting the asymptotic formula $(2N)^n + O(N^{n-1})$, was ultimately established by Bhargava [4]:

Theorem (Bhargava). *If $n \geq 5$, then*

$$|\mathscr{P}^0_{n,N}(\mathbb{Q})| = (2N)^n + O(N^{n-1}) \quad \textit{as } N \to \infty.$$

Chow and Dietmann [5] subsequently proved analogous formulas for the cubic and quartic cases. Theorem 1.1 below extends these quantitative bounds to general number fields K of degree d, providing a unified description of how frequently full symmetric Galois groups arise in families of integral polynomials. The proof combines the classical large-sieve framework of [2] with the recent geometric and analytic refinements developed in [6, 7]. Full technical details can be found in those references; here, we summarize the statement for completeness.

2. Number of S_n-polynomials over K

Theorem 1.1. *Let $d \geq 1$ and $n \geq 2$. There exist constants $\theta > 0$ and $\theta_n \geq 0$ such that the number of non-S_n polynomials satisfies*

$$|\mathscr{P}_{n,N}(K) \setminus \mathscr{P}^0_{n,N}(K)| \ll_{n,K} N^{d(n-\theta)} (\log N)^{\theta_n},$$

as $N \to \infty$. In particular:

1. *If $n = 2$, one can take $\theta = 1$ and $\theta_2 = 1$.*
2. *For all $d \geq 1$ and $n \geq 3$, the bound holds with $\theta = \frac{1}{2}$ and $\theta_n = 1 - \gamma_n$, where $\gamma_n \sim (2\pi n)^{-1/2}$.*
3. *In each of the following cases, $\theta = 1$ and $\theta_n = 0$ are admissible:*

- $d = 1$ for all n;
- arbitrary d with $n = 4, 5$;
- any $n \geq 2(2d + 1)$.

Let G be a subgroup of S_n and is defined

$$\mathscr{N}_n(N, G; K) = \{f \in \mathscr{P}_{n,N}(K) : G_f \cong G\}, \qquad N_n(N, G; K) = |\mathscr{N}_n(N, G; K)|.$$

Then Theorem 1.1 asserts that

$$N_n(N, G) \ll_{n,K} N^{d(n-\theta)} (\log N)^{\theta_n} \tag{1}$$

as $N \to \infty$ for every $G \subseteq S_n$.

The first case in part (3) of Theorem 1.1 reproduces van der Waerden's conjecture for polynomials over $\mathbb{Q}$ (proved for specific degrees in [4, 5]), while the remaining bullets extend these asymptotics to number fields K of higher degree and to a broad range of n. The estimates in (2) stem from applying large sieve bounds to the family $\mathscr{P}_{n,N}(K)$, paralleling Gallagher's argument [2] for the rational case. Complete proofs and refinements appear in [6].

3. The number of splitting primes

We now turn to the arithmetic statistics governing the average distribution of splitting types of primes in the extensions generated by polynomials in $\mathscr{P}^0_{n,N}(K)$. The goal is to quantify, on average, how often a given splitting type occurs, in the spirit of a probabilistic Chebotarev theorem.

Fix a square-free splitting type $r = (r_1, \dots, r_n)$ and a real number $x \geq 1$. For $f \in \mathscr{P}^0_{n,N}(K)$ define

$$\pi_{f,r}(x) = \sum_{\substack{\wp \subseteq \mathcal{O}_K \\ N_{K/\mathbb{Q}}\wp \leq x}} 1_{f,r}(\wp) = \sum_{\substack{\wp \subseteq \mathcal{O}_K \\ N_{K/\mathbb{Q}}\wp \leq x}} \begin{cases} 1, & \text{if } f \text{ has splitting type } r \bmod \wp, \\ 0, & \text{otherwise.} \end{cases}$$

Thus, $\pi_{f,r}(x)$ counts the number of unramified primes of norm at most x whose Frobenius conjugacy class in the splitting field K_f corresponds to the cycle structure r.

We endow $\mathscr{P}^0_{n,N}(K)$ with the uniform probability measure $\mathbb{P}_N$ induced by the distribution of coefficients in $[-N, N]^{nd}$. The corresponding expectation and variance operators are denoted $\mathbb{E}_N$ and σ_N^2. If $\wp \subseteq \mathcal{O}_K$ is a prime ideal, write $N_{K/\mathbb{Q}}\wp = p_\wp^{f_\wp} = q_\wp$. Let $\pi_K(x)$ count primes of $\mathcal{O}_K$ with norm at most x; by the prime ideal theorem, $\pi_K(x) \sim \mathrm{Li}(x)$.

The next statement describes the asymptotic distribution of $\pi_{f,r}(x)$ for large families of polynomials.

Theorem 1. *For $x = N^{1/\log\log N}$ and real numbers $a < b$, one has*

$$\mathbb{P}_N\left(a \leq \frac{\pi_{f,r}(x) - \delta(r)\pi_K(x)}{\left(\delta(r) - \delta(r)^2\right)^{1/2} \pi_K(x)^{1/2}} \leq b \right) \to \frac{1}{\sqrt{2\pi}} \int_a^b e^{-t^2/2} dt \quad \text{as } N \to \infty,$$

where $\delta(r)$ is the proportion of elements in S_n having cycle type r.

This central limit theorem shows that, for almost all polynomials in the family, the normalized count of primes with a fixed splitting type fluctuates according to a Gaussian law centered at the Chebotarev-predicted mean $\delta(r)\pi_K(x)$.

4. An explicit example

4.1 Splitting types of cubic polynomials over $\mathbb{Q}$

We start with a worked-out example with explicit computations of splitting types for a small-degree polynomial and a short numerical experiment illustrating the predictions of Chebotarev's theorem on average splitting density.

Consider the family of monic cubic polynomials over $\mathbb{Z}$ of the form $f(x) = x^3 + ax + b$, with integral coefficients a, b bounded by some height N. For simplicity, take $N = 10$. For a fixed prime (p), the polynomial f modulo (p) can factor into irreducible factors of degrees partitioning 3. The possible splitting types correspond to cycle types of S_3:

- (3): An irreducible cubic mod (p) (cycle type: 3-cycle).
- $(2, 1)$: A quadratic irreducible factor and a linear factor.
- (1,1,1): Completely split into three distinct linear factors.

By Chebotarev's Density Theorem, the density of primes with a given splitting type corresponds to the proportion of elements in the Galois group S_3 with the associated cycle structure, as one can see in **Table 1**.

4.1.1 Explicit computation for a sample polynomial

Take $f(x) = x^3 - 2x + 1$, which is irreducible over $\mathbb{Q}$ with Galois group S_3. Check splitting mod primes $p = 3,5,7,11$:

- $(p = 3)$: $f(x) \equiv x^3 - 2x + 1 \equiv x^3 + x + 1 \pmod{3}$. Factorization in $\mathbb{F}_3[x]$: try $(x = 0,1,2)$, none is a root (since $(1,0,2 \neq 0 \mod 3)$, so f is irreducible mod 3 $\rightarrow$ splitting type (3).
- $(p = 5)$: $f(x) \equiv x^3 - 2x + 1 \equiv x^3 + 3x + 1 \pmod{5}$. Check roots mod 5: $x = 1$, $1 + 3 + 1 = 5 \equiv 0$, root found $\rightarrow$ linear factor. Polynomial divides as $(x - 1)(x^2 + \dots) \rightarrow$ splitting type $(2, 1)$.
- $(p = 7)$: $f(x) \equiv x^3 - 2x + 1 \pmod{7}$. Test roots $x = 0, 1, \dots, 6$: none zero, so irreducible $\rightarrow$ splitting type (3).
- $(p = 11)$: $f(x) \equiv x^3 - 2x + 1 \pmod{11}$ Check roots: $x = 1$, $1 - 2 + 1 = 0$, root $\rightarrow$ splitting type $(2, 1)$.

4.1.2 Numerical experiment: Empirical density of splitting types

For primes $p \leq 31$, compute the splitting types of $f(x) = x^3 - 2x + 1 \mod p$: Count of types:

- Fully split (1,1,1: 2/11≈18%
- Split $(2, 1)$: 6/11≈55%
- Irreducible (3) : 3/11≈27%

Splitting type	Cycle type in S_3	Proportion $\delta(r)$
(1,1,1) (fully split)	Identity (1)(2)(3)	1/6
(2,1) (linear + quadratic)	Transpositions (3 elements)	$3/6 = 1/2$
(3) (irreducible)	3-cycles (2 elements)	$2/6 = 1/3$

Table 1.
Possible splitting types of a cubic polynomial mod primes and their theoretical densities according to Chebotarev's theorem with Galois group S_3.

Prime p	Splitting type
2	(2,1)
3	(3)
5	(2,1)
7	(3)
11	(2,1)
13	(2,1)
17	(1,1,1)
19	(2,1)
23	(1,1,1)
29	(3)
31	(2,1)

Table 2.
Splitting types of $f(x) = x^3 - 2x + 1$ modulo primes up to 31.

The theoretical densities for S_3 were $1/6 \approx 16.7\%$ (fully split), $1/2 = 50\%$ (2,1), and $1/3 \approx 33.3\%$ (irreducible). Our small sample aligns reasonably well with these predicted proportions even with this very limited data set (**Table 2**).

This example explicitly illustrates the connection between factorization types of f modulo primes, the cycle structure in the Galois group, and the statistical predictions of Chebotarev's theorem. Extending this to large sets of polynomials or larger prime ranges recovers the average behaviors studied in this paper.

5. Effective Chebotarev density theorem on average

We next derive an averaged, quantitative version of the Chebotarev density law within the family $\mathscr{P}^0_{n,N}(K)$. Let $\xi > 0$ be a constant such that

$$|\mathscr{P}_{n,N}(K) \backslash \mathscr{P}^0_{n,N}(K)| \ll_{n,K} N^{d(n-\xi)}.$$

For general $n \geq 3$ and $d \geq 1$ one may take $\xi = \frac{1}{2} - \varepsilon$, while in the special settings listed in part (3) of Theorem 1.1 one can choose $\xi = 1$.

For a splitting type r and a prime $\wp$ of norm $q_\wp$, denote by $X_{n,r,\wp}$ the set of monic degree-n polynomials over $\mathbb{F}_{q_\wp}$ having square-free factorization pattern r. The following lemma, adapted from [7], gives the essential counting estimate for congruence conditions across several primes.

Lemma 4.1. *Let $k \geq 1$, and let $\wp_1, \ldots, \wp_k$ be distinct primes with corresponding elements $g_i \in X_{n,r,\wp_i}$ for $i = 1, \ldots, k$. If $q_{\wp_i} < N^{d\xi/(kn)}$ for all i, then*

$$\mathbb{P}_N\left(f \in \mathscr{P}^0_{n,N} : f \equiv g_i \pmod{\wp_i} \text{ for all } i\right) = \frac{1}{\left(q_{\wp_1} \cdots q_{\wp_k}\right)^n} + O_{n,K}\left(N^{-d\xi}\right) \quad \text{as } N \to \infty.$$

This approximation asserts that local conditions at small primes behave almost independently when N is large, with an error controlled by $N^{-d\xi}$.

Proposition 4.1. *For every prime* $\wp$ *satisfying* $q_\wp < N^{d\xi/(n+1)}$, *the following hold:*

1. $\mathbb{P}_N(1_{f,r}(\wp) = 1) = \mathbb{E}_N(1_{f,r}(\wp)) = \delta(r) + \frac{C_r}{q_\wp} + O\left(\frac{1}{q_\wp^2} + q_\wp^n N^{-d\xi}\right)$, for some constant C_r depending only on r;

2. $\sigma_N^2(1_{f,r}(\wp)) = \left(\delta(r) - \delta(r)^2\right) + \frac{C_r(1-2\delta(r))}{q_\wp} + O\left(\frac{1}{q_\wp^2} + q_\wp^n N^{-d\xi}\right)$;

3. *for* $x < N^{d\xi/(n+1)}$,

$$\mathbb{E}_N(\pi_{f,r}(x)) = \delta(r)\pi_K(x) + C_r \log\log x + O_{n,K}(1).$$

Hence, the expected number of primes of norm up to x with splitting type r equals $\delta(r)\pi_K(x)$ to leading order, matching the classical Chebotarev density, with secondary fluctuations of logarithmic size.

Remark. From part (3) above, for any integer $m \geq 2$ one deduces that

$$\pi_{f,r}(x) - \delta(r)\pi_K(x) = O((\log\log x)^m) \quad \text{as } x \to \infty,$$

for all but $O_{n,K}\left(x^{\frac{(n+1)(n-\xi)}{\xi}} (\log\log x)^{1-m}\right)$ polynomials f with height $\ll x^{(n+1)/(d\xi)}$.

Let $\varphi : G_f \to \mathbb{C}$ be any class function. Define

$$\pi_{f,\varphi}(x) = \sum_{\substack{N_{K/\mathbb{Q}}\wp \leq x \\ \wp \nmid D_f}} \varphi(\mathrm{Frob}_{f,\wp}) = \sum_r \varphi(g_r)\pi_{f,r}(x),$$

where g_r represents the conjugacy class $\mathscr{C}_r$. Averaging over $\mathscr{P}^0_{n,N}(K)$ gives the following consequence.

Corollary 4.1. *If* $x < N^{d\xi/(n+1)}$, *then*

$$\mathbb{E}_N(\pi_{f,\varphi}(x)) = \sum_r \delta(r)\varphi(g_r)\pi_K(x) + \sum_r \delta(r)\varphi(g_r)\log\log x + O_{n,K}(\|\varphi\|),$$

where $\|\varphi\| = \sup_{g\in G_f} |\varphi(g)|$.

6. Higher moments

To establish Theorem 1, we employ the method of moments, adapting ideas of Granville and Soundararajan [8]. For each prime $\wp$, define the Bernoulli random variable

$$X_\wp = 1_{f,r}(\wp), \qquad \mathbb{P}(X_\wp = 1) = \frac{|X_{n,r,\wp}|}{q_\wp^n} = \delta(r) + \frac{C_r}{q_\wp} + O\left(q_\wp^{-2}\right).$$

Let

$$Y_\wp(f) = \begin{cases} 1 - \dfrac{|X_{n,r,\wp}|}{q_\wp^n}, & \text{if } 1_{f,r}(\wp) = 1, \\ -\dfrac{|X_{n,r,\wp}|}{q_\wp^n}, & \text{otherwise.} \end{cases}$$

For an ideal $\mathfrak{a} = \prod_i \wp_i^{\alpha_i}$ of $\mathcal{O}_K$, put $Y_{\mathfrak{a}}(f) = \prod_i Y_{\wp_i}(f)^{\alpha_i}$.
Lemma 5.1. *Uniformly for even integers k with* $k \ll_r \frac{d\xi}{n+1/2}\frac{\log N}{\log z}$, *one has*

$$\frac{1}{|\mathscr{P}^0_{n,N}|}\sum_{f\in\mathscr{P}^0_{n,N}}\left(\sum_{N_{K/\mathbb{Q}}\wp\le z} Y_{\wp}(f)\right)^k = C_{k,r}\pi_K(z)^{k/2}\left(1+O\left(\frac{k^3}{(1-\delta(r))^{k/2}}\frac{\log\log z}{\pi_K(z)}\right)\right)$$
$$+O\left(\pi_K(z)^{k(n+1)}N^{-d\xi}\right),$$

and for odd k in the same range,

$$\frac{1}{|\mathscr{P}^0_{n,N}|}\sum_{f\in\mathscr{P}^0_{n,N}}\left(\sum_{N_{K/\mathbb{Q}}\wp\le z} Y_{\wp}(f)\right)^k \ll C_{k,r}\pi_K(z)^{k/2}k\frac{\log\log z}{\pi_K(z)^{1/2}} + \pi_K(z)^{k(n+1)}N^{-d\xi}.$$

Here

$$C_k = \begin{cases}\dfrac{k!}{2^{k/2}(k/2)!}, & k \text{ even},\\[2mm] \dfrac{k!}{2^{(k-1)/2}((k-1)/2)!}, & k \text{ odd},\end{cases}\qquad C_{k,r} = \begin{cases} C_k\left(\delta(r)-\delta(r)^2\right)^{k/2}, & k \text{ even},\\ C_k\delta(r)^{(k-1)/2}, & k \text{ odd}.\end{cases}$$

Proceeding as in [7], one derives:
Proposition 5.1. *Uniformly for even k with* $k \ll_r \frac{d\xi}{n+1/2}\frac{\log N}{\log x}$,

$$\frac{1}{|\mathscr{P}^0_{n,N}|}\sum_{f\in\mathscr{P}^0_{n,N}}\left(\pi_{f,r}(x)-\delta(r)\pi_K(x)\right)^k = C_{k,r}\pi_K(x)^{k/2}\left(1+O\left(\frac{k^{3/2}}{(1-\delta(r))^{k/2}}\frac{\log\log x}{\pi_K(x)^{1/2}}\right)\right)$$
$$+O\left(\pi_K(x)^{k(n+1)}N^{-d\xi}\right),$$

and for odd k in the same range,

$$\frac{1}{|\mathscr{P}^0_{n,N}|}\sum_{f\in\mathscr{P}^0_{n,N}}\left(\pi_{f,r}(x)-\delta(r)\pi_K(x)\right)^k \ll C_{k,r}\pi_K(x)^{k/2}k\frac{\log\log x}{\pi_K(x)^{1/2}} + \pi_K(x)^{k(n+1)}N^{-d\xi}.$$

For $x = o\left(N^{d\xi/[k(n+1/2)]}\right)$, the final error term is negligible, leading to convergence of normalized moments to those of the standard Gaussian. Consequently, the random variable

$$\frac{\pi_{f,r}(x)-\delta(r)\pi_K(x)}{\left(\delta(r)-\delta(r)^2\right)^{1/2}\pi_K(x)^{1/2}}$$

converges in distribution to $\mathcal{N}(0,1)$ as $N\to\infty$, proving Theorem 1.

7. Estimates for subfamilies

We finally extend the previous probabilistic framework to structured subfamilies of $\mathscr{P}^0_{n,N}(K)$, defined by auxiliary parameters that restrict coefficients or impose additional algebraic conditions. This allows one to isolate smaller collections of polynomials while preserving the same limiting statistical behavior.

Let $\mathscr{A} = \mathscr{A}_{\beta,M}(N) \subseteq \mathscr{P}^0_{n,N}(K)$ depend on parameters $\beta > 1$ and $M > 0$, and assume the following hypotheses hold for some fixed splitting type r and positive real x:

1. $1 < \beta \leq n$ and $M > 0$.

2. For each prime $\wp$, define

$$\mathscr{A}^{\wp} = \{f \bmod \wp : f \in \mathscr{A}\} \subseteq \mathbb{F}_{q_\wp}[X].$$

If $\wp_1, \ldots, \wp_s$ are distinct primes with $N_{K/\mathbb{Q}}\wp_i \leq x$ and $g_i \in \mathscr{A}^{\wp_i}$, then

$$\sum_{\substack{f \in \mathscr{A} \\ f \equiv g_i \mod \wp_i \ \forall i}} 1 = \frac{MN^{d\beta}}{\left(q_{\wp_1} \cdots q_{\wp_s}\right)^{\beta}} + E_{n,K}(N),$$

uniformly in the $\wp_i$, where $E_{n,K}(N) \ll N^{d(\beta - \xi)}$ for large N and small x relative to N.

3. Let $X^{\mathscr{A}}_{n,r,\wp} = X_{n,r,\wp} \cap \mathscr{A}^{\wp}$. Suppose that

$$|X^{\mathscr{A}}_{n,r,\wp}| = \delta(r) q^{\beta}_{\wp} + O(E_\wp),$$

where typically $E_\wp \ll q_\wp^{\beta - 1/2}$ when the Riemann hypothesis for curves over finite fields applies. Thus $|X^{\mathscr{A}}_{n,r,\wp}|$ behaves like $\delta(r)|\mathscr{A}^{\wp}|$ with a square-root cancelation error term.

Define, for $f \in \mathscr{A}$,

$$Y_\wp(f) = \begin{cases} 1 - \dfrac{|X^{\mathscr{A}}_{n,r,\wp}|}{q^{\beta}_{\wp}}, & \text{if } \mathbb{1}_{f,r}(\wp) = 1, \\ -\dfrac{|X^{\mathscr{A}}_{n,r,\wp}|}{q^{\beta}_{\wp}}, & \text{otherwise.} \end{cases}$$

Let

$$\mu(x) = \sum_{N_{K/\mathbb{Q}}\wp \leq x} \frac{|X^{\mathscr{A}}_{n,r,\wp}|}{q^{\beta}_{\wp}} = \delta(r)\pi_K(x) + O\left(\frac{\sqrt{x}}{\log x}\right).$$

Then, for each $f \in \mathscr{A}$,

$$\pi_{f,r}(x) - \mu(x) = \sum_{N_{K/\mathbb{Q}}\wp \leq x} Y_\wp(f).$$

Moments of these fluctuations can again be evaluated by expanding over ideals $\mathfrak{a} = \prod_i \wp_i^{\alpha_i}$ of $\mathcal{O}_K$ and setting $Y_{\mathfrak{a}}(f) = \prod_i Y_{\wp_i}(f)^{\alpha_i}$.

Proposition 6.1. *Let* $\mathcal{A} \subseteq \mathcal{P}^0_{n,N}(K)$ *satisfy conditions* (1)–(3).

1. *For even integers k with* $k \ll_{r,n} \frac{\log(MN^{d\beta})}{\log x}$,

$$\sum_{f\in\mathcal{A}} \left(\pi_{f,r}(x) - \mu(x)\right)^k = C_{k,r} MN^{d\beta} \pi_K(x)^{k/2} \left(1 + O\left(\frac{k^3}{(1-\delta(r))^{k/2}} \frac{\sqrt{x}}{\pi_K(x)\log x}\right)\right)$$
$$+O\left(\pi_K(x)^{k(n+1)} E_{n,K}(N)\right),$$

as $x, N \to \infty$.

2. *For odd k with* $k \ll_{r,n} \min\left((\log x)^{1/2}, \frac{\log(MN^{d\beta})}{\log x}\right)$,

$$\sum_{f\in\mathcal{A}} \left(\pi_{f,r}(x) - \mu(x)\right)^k \ll C_{k,r} MN^{d\beta} \pi_K(x)^{k/2} k \frac{\sqrt{x}}{\pi_K(x)^{1/2}\log x} + \pi_K(x)^{k(n+1)} E_{n,K}(N),$$

as $x, N \to \infty$.

3. *For fixed even k, if* $x = o\left(\left(MN^{d\xi}\right)^{1/\left[k\left(n+\frac{1}{2}\right)\right]}\right)$, *then*

$$\sum_{f\in\mathcal{A}} \left(\pi_{f,r}(x) - \mu(x)\right)^k = C_{k,r} MN^{d\beta} \pi_K(x)^{k/2} \left(1 + O_k\left(\frac{\sqrt{x}}{\pi_K(x)\log x}\right)\right)$$
$$+O\left(\pi_K(x)^{k(n+1)} E_{n,K}(N)\right),$$

as $x, N \to \infty$.

The argument mirrors the one developed for the full family, with modifications to account for the weighting factor $MN^{d\beta}$ and the slightly stronger error control afforded by the hypotheses on $\mathcal{A}$.

Note. Applying Proposition 6.1 with a small adjustment of the truncation parameter z leads to a refined bound for the even moments.

Corollary 6.1. *Let* $\mathcal{A} \subseteq \mathcal{P}^0_{n,N}(K)$ *satisfy* (1)–(3). *Then uniformly for even k with* $k \ll_{r,n} \min\left((\log x)^{1/2}, \frac{\log(MN^{d\beta})}{\log x}\right)$, *one has*

$$\sum_{f\in\mathcal{A}} \left(\pi_{f,r}(x) - \mu(x)\right)^k = C_{k,r} MN^{d\beta} \pi_K(x)^{k/2} \left(1 + O\left(\frac{k^{3/2}}{(1-\delta(r))^{k/2}} \frac{\sqrt{x}}{\pi_K(x)^{1/2}\log x}\right)\right)$$
$$+ O\left(\pi_K(x)^{k(n+1)} E_{n,K}(N)\right),$$

as $x, N \to \infty$.

These results demonstrate that the Gaussian fluctuations described by Theorem 1 persist uniformly across a broad spectrum of subfamilies, provided they exhibit sufficient independence and density with respect to the ambient box $[-N, N]^{nd}$.

DOI: http://dx.doi.org/10.5772/intechopen.1013824

Author details

Ilaria Viglino
Institut de Mathématiques, EPFL École polytechnique fédérale de Lausanne, Switzerland

*Address all correspondence to: ilaria.viglino@yahoo.it

IntechOpen

© 2026 The Author(s). Licensee IntechOpen. This chapter is distributed under the terms of the Creative Commons Attribution License (http://creativecommons.org/licenses/by/4.0), which permits unrestricted use, distribution, and reproduction in any medium, provided the original work is properly cited.

References

[1] van der Waerden RJ. Die Seltenheit der reduziblen Gleichungen und der Gleichungenmit Affekt. Monatshefte Mathematical Physics. 1936;**43**(1): 133-147

[2] Gallagher PX. The large sieve and probabilistic Galois theory. In Analytic number theory (Proceedings of Symposia in Pure Mathematics., Vol. XXIV, St. Louis University, St. Louis, Missouri, 1972). Providence, Rhode Island: American Mathematical Society; 1973. pp. 91-101

[3] Dietmann R. Probabilistic Galois theory. The Bulletin of the London Mathematical Society. 2013;**45**(3): 453-462

[4] Bhargava M. Galois groups of random integer polynomials and van der Waerden's conjecture. arXiv: 2111.06507v1 [math.NT] 12 Nov 2021

[5] Chow S, Dietmann R. Enumerative Galois theory for cubics and quartics. Advances in Mathematics. 2020;**372**: 107282

[6] Viglino I. Towards a generalization of the van der Waerden's conjecture for S_n-polynomials with integral coefficients over a fixed number field extension arXiv:2212.11608 [math.NT]

[7] Viglino I. Arithmetic statistics of families of integer S_n-polynomials and application to class group torsion. Journal of Number Theory. 2024;**271**: 33-65. DOI: 10.1016/j.jnt.2024.10.009

[8] Granville A, Soundararajan K. Sieving and the Erdős-Kac theorem. In: Granville A, Rudnick Z, editors. Equidistribution in Number Theory, an Introduction. NATO Science Series. Vol. 237. Dordrecht: Springer; 2007

Chapter 2

Perspective Chapter: Experimental Insights on Prime Gaps

Mircea Ghidarcea

Abstract

Advanced sieving techniques can be used to generate extensive datasets of prime gaps, leading to novel insights and conjectures concerning their distribution. By leveraging the efficiency of refined sieving algorithms, we have systematically analyzed the distribution patterns of prime gaps within large intervals. Our findings include significant statistical trends and potential conjectures regarding the behavior of these gaps, which could provide a deeper understanding of number theory. This research highlights the effectiveness of sieving methods in large-scale prime number analysis and opens new avenues for further mathematical investigation and algorithmic optimization.

Keywords: prime gaps, prime numbers sieving, prime gaps distribution, tandem gaps, number theory

1. Introduction

Prime numbers, the building blocks of arithmetic, have fascinated mathematicians for centuries due to their intrinsic properties and patterns. One particularly intriguing aspect of prime numbers is the distribution of gaps between consecutive primes. Understanding these gaps not only deepens our comprehension of the nature of primes but also has implications in fields such as cryptography, random number generation, and computational mathematics.

Historically, the study of prime gaps can be traced back to the work of Euclid, who proved the infinitude of primes around 300 BCE, implicitly highlighting the nonconstant nature of gaps between primes. The systematic study of these gaps, however, began in earnest in the nineteenth century with Carl Friedrich Gauss and Adrien-Marie Legendre, who independently conjectured the Prime Number Theorem. This theorem, later proven by Hadamard and de la Vallée Poussin in 1896, suggests that the average gap between primes near a large number n is approximately $\log(n)$ [1, 2].

In the twentieth century, significant advancements were made by mathematicians such as Paul Erdős and Atle Selberg. Erdős introduced probabilistic methods to number theory, resulting in profound insights into the distribution of prime gaps [3]. Selberg's foundational work on the distribution of prime numbers, including his elementary proof of the Prime Number Theorem, laid the groundwork for later investigations into prime gaps in arithmetic progressions [4].

Recent research has continued to push the boundaries of our understanding. The landmark work by Yitang Zhang in 2013, which demonstrated the existence of infinitely many pairs of primes with gaps smaller than 70 million, marked a significant breakthrough. This result, coupled with subsequent refinements by James Maynard and the Polymath Project, has significantly narrowed the bounds on prime gaps, highlighting the dynamic and ongoing nature of this research area [5, 6].

The mechanisms to generate consecutive primes and prime gaps are based on sieving techniques and algorithms—the status of the sieving domain is analyzed in Refs. [7, 8]. Advanced sieving techniques have made it feasible to generate and analyze large sets of prime numbers efficiently—some of the latest advances in sieving algorithms are presented in Refs. [9–12]. These methods, building upon the classic algorithms such as the Sieve of Eratosthenes or the Sieve of Atkin, allow us to explore the prime gaps in unprecedented detail. The improvements in algorithmic efficiency and computational power have opened new avenues for systematic studies of prime gaps, leading to novel insights and conjectures.

In this chapter, after we employed refined sieving techniques to generate extensive datasets of prime gaps, we analyzed their distribution patterns across large numerical intervals. Our approach not only confirms known statistical trends but also uncovers new patterns and conjectures regarding the behavior of prime gaps. By leveraging the computational efficiency of modern sieving algorithms, we aim to provide a deeper understanding of the distribution of prime gaps and contribute to the ongoing discourse in number theory.

The significance of this research lies in its potential applications and theoretical implications. The findings from our analysis could inform the design of more efficient algorithms for prime number generation, improve cryptographic systems, and enhance pseudo-random number generators. Moreover, the conjectures proposed in this study could guide future mathematical investigations, leading to a more comprehensive theory of prime gaps.

This introduction sets the stage for a detailed exploration of prime gaps in discrete intervals using advanced sieving techniques, highlighting the importance of efficient algorithmic approaches and the potential for significant mathematical discoveries.

Note 1: This chapter is not concerned directly with sieving techniques— one can find extensive sieving examples in the Reference section. Moreover, we encourage the reader to create their own tools and datasets to further confirm independently the results presented here. The detailed methodology behind the optimized prime sieving algorithm used in this study is similar to the one described in our previous work [9], where we revisit and improve the classical Sieve of Atkin. That publication outlines the algorithmic structure, optimizations, complexity, and empirical performance evaluation. In this study, we employ a similar implementation—available in full at our GitHub repository, named in Appendix A, where we give all the details about our own tools and techniques used for sieving, constructing the datasets, and analyzing them.

2. Prime gaps and PRNGs

As the sequential prime numbers generated through sieving techniques are, for now, too small to be used in cryptography, there is limited practical application for sieving beyond fundamental research in pure computer science algorithmic endeavors.

There are, however, at least two areas of interest, somehow related:

- theoretical studies concerned with (analytic) number theory, like primes or prime gaps distribution, validation of various conjectures, etc.

- pseudo-random number generators (PRNG) or noise generators based on prime number sequences.

Although our background is primarily in computer science, where the focus is more on algorithmic applications such as pseudo-random number generation (PRNG), our investigation into prime-based generators has naturally led us into the realm of number theory. Initially, we were less concerned with theoretical aspects and more with the practical construction of efficient PRNGs. However, we soon discovered that understanding the distribution of primes and prime gaps—central topics in (analytic) number theory—is essential for designing robust prime-based generators. This unexpected overlap has led us to explore the statistical behavior of prime gaps more deeply, using large-scale sieving techniques.

A PRNG is based on those characteristics of the prime distribution that are the subject of areas like number theory or, more particularly, analytic number theory. To validate various hypotheses, one needs to generate as fast as possible large sequences of primes as huge as possible—using sieving techniques, it is possible to explore very large sets of primes. These findings, though rooted in computational experimentation, point toward patterns and regularities that may also hold theoretical significance.

A specific type of PRNG can be constructed based on prime gaps: Prime numbers within a certain interval are generated, and the resulting sequence of gaps is used as random numbers after halving (since all gaps are even: Gap 1 corresponds to a gap of length 2 between twin primes, gap 2 has length 4, and so on). A seed can determine the starting point of the interval to be processed, and an optional reordering of the gap sequence can be applied to ensure uniqueness. However, this study does not focus on pseudo-random number generators. PRNGs are mentioned here solely to provide context and to illustrate one possible application that motivated our interest in analyzing prime gaps. We will not elaborate on sieve or PRNG-related implementation details, except where strictly necessary to clarify the relevance of certain statistical properties of the gaps. Understanding the distribution of prime gaps is a prerequisite for such applications—but the core of this work remains an empirical investigation into the structure and behavior of these gaps themselves.

Note 2: To simplify representation, all prime gaps shown in tables and figures use halved values. Since prime gaps are always even, this convention makes the data more compact while preserving all structural properties. In practice, using halved values, actual gaps up to 512 can be encoded using just one byte.

Figure 1 is an example of such a study: the distribution/frequency of prime gaps, analyzed on sets of 150 million primes using an absolute step of 1 million. **Figures 1(a)** and **(b)** display the 3D distribution of prime gap frequencies across consecutive sets of 150 million prime numbers. The starting point for each set is offset in absolute value by 1,000,000 relative to the previous one. For instance, the first set includes the first 150 million primes greater than or equal to 0, the second set starts at 1,000,000 and includes the next 150 million primes from that point, the third set begins at 2,000,000, and so on. This allows for a detailed analysis of how prime gap distributions evolve across increasing numeric intervals.

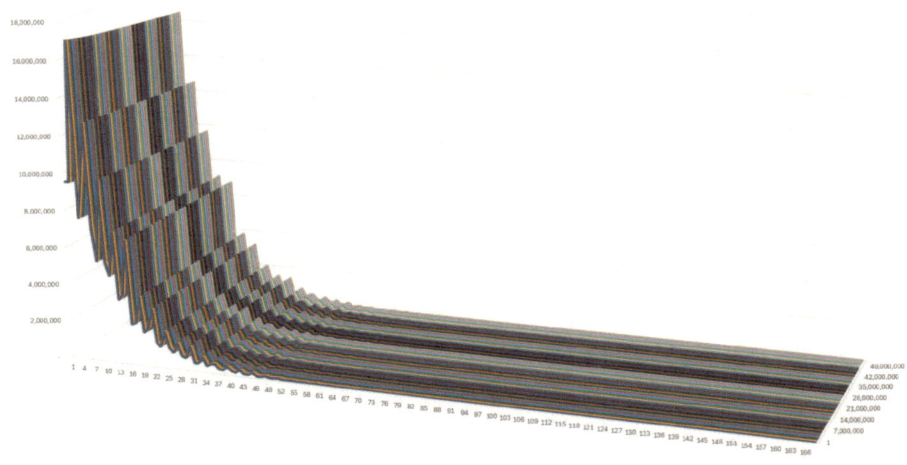

(a) Normal scale, showing the consistent distribution pattern across all sets

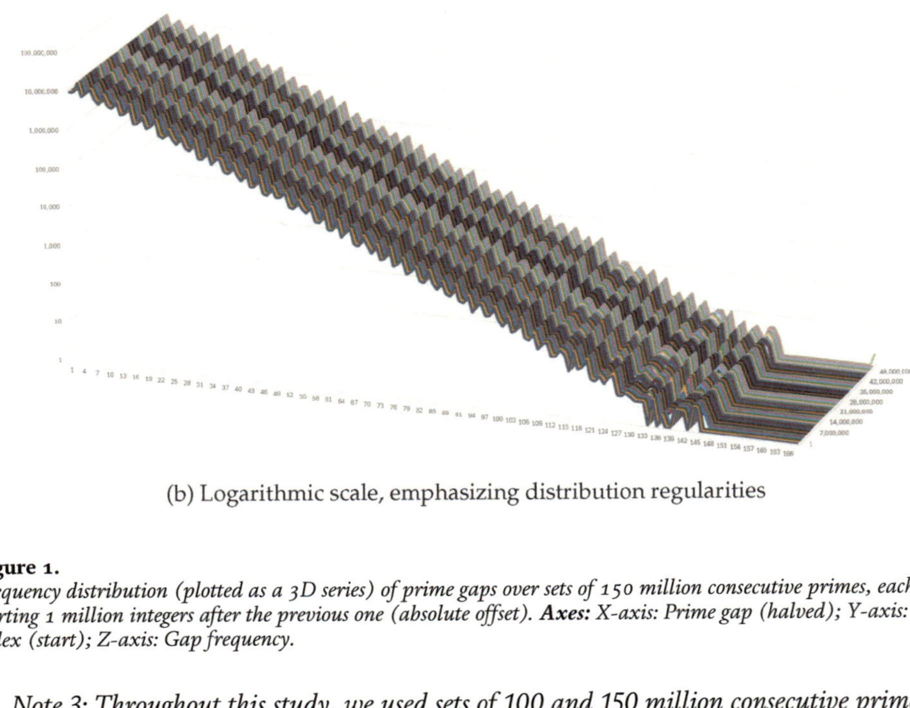

(b) Logarithmic scale, emphasizing distribution regularities

Figure 1.
Frequency distribution (plotted as a 3D series) of prime gaps over sets of 150 million consecutive primes, each set starting 1 million integers after the previous one (absolute offset). ***Axes:*** *X-axis: Prime gap (halved); Y-axis: Set index (start); Z-axis: Gap frequency.*

Note 3: Throughout this study, we used sets of 100 and 150 million consecutive primes, chosen as a practical compromise between statistical significance and computational feasibility. A step of 1 million primes was used between starting points of each set, providing sufficient granularity for observing distributional trends across overlapping intervals.

The figure is significant, showing a relatively consistent and similar pattern for the distribution of prime gaps, regardless of the interval in which they are analyzed. Changing the chart to a logarithmic scale in **Figure 1(b)**, things become even clearer: The gaps approximate a logarithmic distribution, which is basically maintained no matter where the sample of 150 million consecutive primes is taken from.

Using line charts, this character is even more evident—practically the lines overimpose one another, as seen in **Figure 2(a)**. Using a logarithmic scale—**Figure 2(b)**—

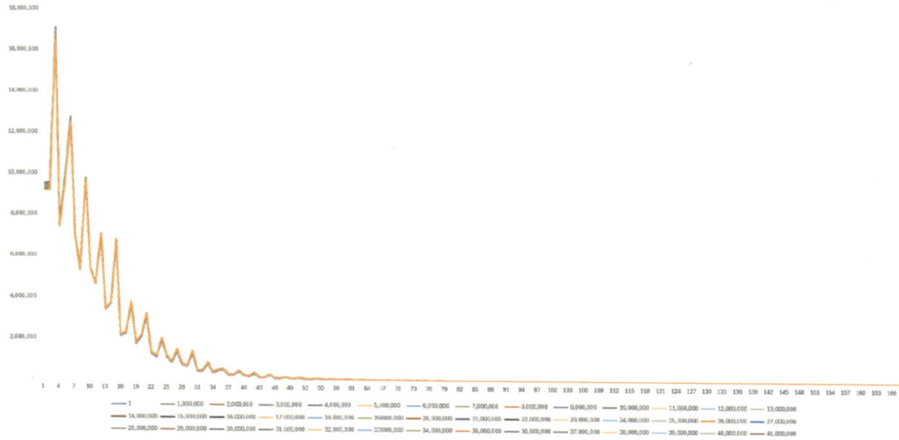

(a) Normal scale

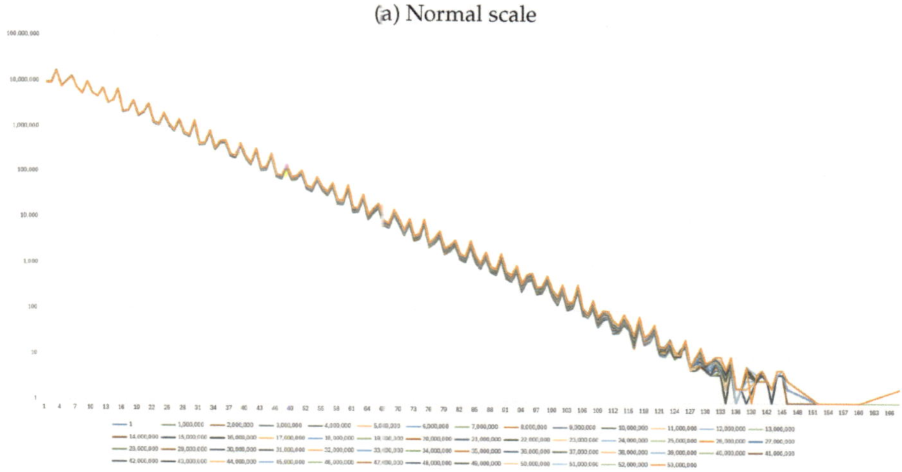

(b) Logarithmic scale

Figure 2.
Same frequency distribution, plotted as 2D lines.

will expose some minor differences for larger gaps, but those are less and less significant as the primes are larger. The logarithmic representation reveals that the gaps align along an almost linear trajectory, with a slowly diminishing slope as the prime intervals shift toward higher absolute values.

Thus, one can get a pretty good PRNG for a logarithmic distribution, as long as the range 1–150 is enough—the range can be greater for larger primes. This type of random number sequence can be useful in cases such as Monte Carlo simulations, statistical sampling for surveys, randomized control trials, or in machine learning and artificial intelligence for data shuffling during training or initialization of neural network weights, particularly when a logarithmic distribution is appropriate.

Further insights can be drawn from these distributions. **Figure 3** presents the normalized frequencies (expressed as percentages relative to the total number of primes in each sequence) for the first 35 gaps, based on the first scenario: sequences of 100 million consecutive primes. As previously noted, each sequence begins at an absolute offset of 1 million from the previous one. Each line in the table corresponds to

Start	1	2	3	4	5	6	7	8	9	10	11	12	13	14	15	16	17	18	19	20	21	22	23	24	25	26	27	28	29	30	31	32	33	34	35
0	6.50	6.50	11.60	5.16	6.68	8.62	4.77	3.59	6.54	3.59	3.08	4.68	2.23	2.43	4.38	1.39	1.46	2.39	1.12	1.33	1.98	0.81	0.70	1.21	0.69	0.52	0.87	0.45	0.39	0.81	0.25	0.26	0.48	0.20	0.28
1000000	6.47	6.47	11.57	5.14	6.67	8.61	4.77	3.59	6.54	3.59	3.09	4.69	2.23	2.44	4.39	1.39	1.46	2.39	1.13	1.34	1.99	0.82	0.70	1.22	0.70	0.52	0.88	0.46	0.39	0.81	0.25	0.26	0.48	0.20	0.28
2000000	6.46	6.46	11.54	5.14	6.66	8.60	4.76	3.58	6.54	3.60	3.09	4.69	2.24	2.44	4.40	1.39	1.46	2.40	1.13	1.34	2.00	0.82	0.71	1.23	0.70	0.53	0.88	0.46	0.40	0.82	0.26	0.26	0.48	0.20	0.28
3000000	6.44	6.44	11.52	5.13	6.65	8.59	4.76	3.58	6.54	3.60	3.09	4.70	2.24	2.44	4.40	1.39	1.47	2.41	1.14	1.35	2.00	0.82	0.71	1.23	0.70	0.53	0.89	0.46	0.40	0.82	0.26	0.26	0.49	0.20	0.28
4000000	6.43	6.43	11.50	5.12	6.64	8.58	4.75	3.58	6.53	3.60	3.09	4.70	2.24	2.44	4.41	1.40	1.47	2.41	1.14	1.35	2.01	0.82	0.71	1.23	0.70	0.53	0.89	0.46	0.40	0.82	0.26	0.27	0.49	0.20	0.29
5000000	6.42	6.42	11.48	5.11	6.63	8.57	4.75	3.58	6.53	3.60	3.09	4.70	2.24	2.44	4.41	1.40	1.47	2.41	1.14	1.35	2.01	0.83	0.71	1.24	0.71	0.53	0.89	0.47	0.40	0.83	0.26	0.27	0.49	0.20	0.29
6000000	6.41	6.41	11.47	5.11	6.62	8.56	4.75	3.58	6.53	3.60	3.09	4.70	2.24	2.44	4.42	1.40	1.47	2.42	1.14	1.35	2.02	0.83	0.72	1.24	0.71	0.53	0.90	0.47	0.40	0.83	0.26	0.27	0.49	0.20	0.29
7000000	6.40	6.40	11.45	5.10	6.61	8.56	4.74	3.58	6.53	3.60	3.09	4.70	2.24	2.45	4.42	1.40	1.48	2.42	1.14	1.36	2.02	0.83	0.72	1.25	0.71	0.54	0.90	0.47	0.40	0.83	0.26	0.27	0.50	0.20	0.29
8000000	6.39	6.39	11.44	5.10	6.61	8.55	4.74	3.57	6.53	3.60	3.09	4.70	2.24	2.45	4.43	1.40	1.48	2.42	1.15	1.36	2.02	0.83	0.72	1.25	0.71	0.54	0.90	0.47	0.40	0.84	0.26	0.27	0.50	0.20	0.29
9000000	6.38	6.38	11.43	5.09	6.60	8.54	4.74	3.57	6.52	3.60	3.09	4.70	2.25	2.45	4.43	1.40	1.48	2.43	1.15	1.36	2.03	0.83	0.72	1.25	0.71	0.54	0.91	0.47	0.41	0.84	0.26	0.27	0.50	0.20	0.29
10000000	6.37	6.37	11.41	5.09	6.59	8.54	4.74	3.57	6.52	3.60	3.09	4.70	2.25	2.45	4.43	1.41	1.48	2.43	1.15	1.36	2.03	0.83	0.72	1.25	0.72	0.54	0.91	0.47	0.41	0.84	0.27	0.27	0.50	0.21	0.29
11000000	6.37	6.36	11.40	5.08	6.59	8.53	4.73	3.57	6.52	3.60	3.09	4.70	2.25	2.45	4.44	1.41	1.48	2.43	1.15	1.37	2.03	0.84	0.72	1.26	0.72	0.54	0.91	0.47	0.41	0.84	0.27	0.27	0.50	0.21	0.29
12000000	6.36	6.36	11.39	5.08	6.58	8.53	4.73	3.57	6.52	3.60	3.09	4.71	2.25	2.45	4.44	1.41	1.48	2.44	1.15	1.37	2.04	0.84	0.73	1.26	0.72	0.54	0.91	0.47	0.41	0.85	0.27	0.27	0.50	0.21	0.30
13000000	6.35	6.35	11.38	5.07	6.58	8.52	4.73	3.57	6.52	3.60	3.09	4.71	2.25	2.45	4.44	1.41	1.48	2.44	1.15	1.37	2.04	0.84	0.73	1.26	0.72	0.54	0.91	0.48	0.41	0.85	0.27	0.27	0.51	0.21	0.30
14000000	6.34	6.34	11.37	5.07	6.57	8.52	4.72	3.56	6.52	3.60	3.09	4.71	2.25	2.45	4.44	1.41	1.49	2.44	1.15	1.37	2.04	0.84	0.73	1.26	0.72	0.55	0.92	0.48	0.41	0.85	0.27	0.28	0.51	0.21	0.30
15000000	6.34	6.33	11.35	5.06	6.57	8.51	4.72	3.56	6.51	3.60	3.09	4.71	2.25	2.45	4.45	1.41	1.49	2.44	1.16	1.37	2.05	0.84	0.73	1.27	0.72	0.55	0.92	0.48	0.41	0.85	0.27	0.28	0.51	0.21	0.30
16000000	6.33	6.33	11.34	5.06	6.56	8.51	4.72	3.56	6.51	3.60	3.09	4.71	2.25	2.45	4.45	1.41	1.49	2.45	1.16	1.37	2.05	0.84	0.73	1.27	0.72	0.55	0.92	0.48	0.41	0.86	0.27	0.28	0.51	0.21	0.30
17000000	6.32	6.32	11.34	5.06	6.56	8.50	4.72	3.56	6.51	3.60	3.09	4.71	2.25	2.45	4.45	1.41	1.49	2.45	1.16	1.38	2.05	0.84	0.73	1.27	0.73	0.55	0.92	0.48	0.41	0.86	0.27	0.28	0.51	0.21	0.30
18000000	6.32	6.32	11.33	5.05	6.55	8.50	4.72	3.56	6.51	3.60	3.08	4.71	2.25	2.46	4.45	1.42	1.49	2.45	1.16	1.38	2.06	0.85	0.73	1.27	0.73	0.55	0.92	0.48	0.42	0.86	0.27	0.28	0.51	0.21	0.30
19000000	6.31	6.31	11.32	5.05	6.55	8.49	4.71	3.56	6.51	3.60	3.08	4.71	2.25	2.46	4.46	1.42	1.49	2.45	1.16	1.38	2.06	0.85	0.73	1.27	0.73	0.55	0.93	0.48	0.42	0.86	0.27	0.28	0.52	0.21	0.30
20000000	6.30	6.30	11.31	5.05	6.54	8.49	4.71	3.56	6.51	3.60	3.08	4.71	2.25	2.46	4.46	1.42	1.49	2.45	1.16	1.38	2.06	0.85	0.73	1.28	0.73	0.55	0.93	0.48	0.42	0.86	0.27	0.28	0.52	0.21	0.30
21000000	6.30	6.30	11.30	5.04	6.54	8.48	4.71	3.56	6.50	3.60	3.08	4.71	2.26	2.46	4.46	1.42	1.49	2.46	1.16	1.38	2.06	0.85	0.74	1.28	0.73	0.55	0.93	0.48	0.42	0.87	0.27	0.28	0.52	0.21	0.30
22000000	6.29	6.29	11.29	5.04	6.53	8.48	4.71	3.55	6.50	3.60	3.08	4.71	2.26	2.46	4.46	1.42	1.49	2.46	1.16	1.38	2.07	0.85	0.74	1.28	0.73	0.55	0.93	0.49	0.42	0.87	0.27	0.28	0.52	0.21	0.31
23000000	6.29	6.29	11.28	5.04	6.53	8.47	4.71	3.55	6.50	3.60	3.08	4.71	2.26	2.46	4.47	1.42	1.50	2.46	1.16	1.39	2.07	0.85	0.74	1.28	0.73	0.55	0.93	0.49	0.42	0.87	0.28	0.28	0.52	0.21	0.31
24000000	6.28	6.28	11.27	5.03	6.52	8.47	4.70	3.55	6.50	3.60	3.08	4.71	2.26	2.46	4.47	1.42	1.50	2.46	1.17	1.39	2.07	0.85	0.74	1.28	0.73	0.56	0.93	0.49	0.42	0.87	0.28	0.28	0.52	0.21	0.31
25000000	6.28	6.28	11.26	5.03	6.52	8.47	4.70	3.55	6.50	3.60	3.08	4.71	2.26	2.46	4.47	1.42	1.50	2.46	1.17	1.39	2.07	0.85	0.74	1.29	0.74	0.56	0.94	0.49	0.42	0.87	0.28	0.28	0.52	0.22	0.31
26000000	6.27	6.27	11.25	5.03	6.52	8.46	4.70	3.55	6.50	3.60	3.08	4.71	2.26	2.46	4.47	1.42	1.50	2.46	1.17	1.39	2.08	0.85	0.74	1.29	0.74	0.56	0.94	0.49	0.42	0.88	0.28	0.28	0.52	0.22	0.31
27000000	6.27	6.27	11.25	5.02	6.51	8.46	4.70	3.55	6.50	3.60	3.08	4.71	2.26	2.46	4.47	1.42	1.50	2.47	1.17	1.39	2.08	0.86	0.74	1.29	0.74	0.56	0.94	0.49	0.42	0.88	0.28	0.28	0.53	0.22	0.31
28000000	6.26	6.26	11.24	5.02	6.51	8.45	4.70	3.55	6.49	3.60	3.08	4.71	2.26	2.46	4.47	1.42	1.50	2.47	1.17	1.39	2.08	0.86	0.74	1.29	0.74	0.56	0.94	0.49	0.42	0.88	0.28	0.28	0.53	0.22	0.31
29000000	6.26	6.26	11.23	5.02	6.51	8.45	4.70	3.55	6.49	3.60	3.08	4.71	2.26	2.46	4.48	1.42	1.50	2.47	1.17	1.39	2.08	0.86	0.74	1.29	0.74	0.56	0.94	0.49	0.42	0.88	0.28	0.29	0.53	0.22	0.31
30000000	6.25	6.25	11.22	5.02	6.50	8.45	4.69	3.55	6.49	3.60	3.08	4.72	2.26	2.46	4.48	1.43	1.50	2.47	1.17	1.39	2.08	0.86	0.74	1.29	0.74	0.56	0.94	0.49	0.42	0.88	0.28	0.29	0.53	0.22	0.31
31000000	6.25	6.25	11.22	5.01	6.50	8.44	4.69	3.55	6.49	3.60	3.08	4.72	2.26	2.46	4.48	1.43	1.50	2.47	1.17	1.39	2.09	0.86	0.75	1.30	0.74	0.56	0.94	0.49	0.43	0.88	0.28	0.29	0.53	0.22	0.31
32000000	6.24	6.24	11.21	5.01	6.50	8.44	4.69	3.54	6.49	3.60	3.08	4.72	2.26	2.46	4.48	1.43	1.50	2.48	1.17	1.40	2.09	0.86	0.75	1.30	0.74	0.56	0.95	0.49	0.43	0.89	0.28	0.29	0.53	0.22	0.31
33000000	6.24	6.24	11.20	5.01	6.49	8.44	4.69	3.54	6.49	3.60	3.08	4.72	2.26	2.46	4.48	1.43	1.50	2.48	1.17	1.40	2.09	0.86	0.75	1.30	0.74	0.56	0.95	0.49	0.43	0.89	0.28	0.29	0.53	0.22	0.31
34000000	6.23	6.24	11.19	5.00	6.49	8.43	4.69	3.54	6.49	3.60	3.08	4.72	2.26	2.47	4.48	1.43	1.50	2.48	1.17	1.40	2.09	0.86	0.75	1.30	0.74	0.56	0.95	0.49	0.43	0.89	0.28	0.29	0.53	0.22	0.31
35000000	6.23	6.23	11.19	5.00	6.49	8.43	4.69	3.54	6.49	3.60	3.08	4.72	2.26	2.47	4.49	1.43	1.50	2.48	1.17	1.40	2.09	0.86	0.75	1.30	0.75	0.56	0.95	0.50	0.43	0.89	0.28	0.29	0.53	0.22	0.31
36000000	6.23	6.23	11.18	5.00	6.48	8.43	4.68	3.54	6.49	3.60	3.08	4.72	2.26	2.47	4.49	1.43	1.51	2.48	1.18	1.40	2.09	0.86	0.75	1.30	0.75	0.57	0.95	0.50	0.43	0.89	0.28	0.29	0.54	0.22	0.32
37000000	6.22	6.22	11.17	5.00	6.48	8.42	4.68	3.54	6.48	3.60	3.08	4.72	2.26	2.47	4.49	1.43	1.51	2.48	1.18	1.40	2.10	0.86	0.75	1.30	0.75	0.57	0.95	0.50	0.43	0.89	0.28	0.29	0.54	0.22	0.32
38000000	6.22	6.22	11.17	4.99	6.48	8.42	4.68	3.54	6.48	3.60	3.08	4.72	2.26	2.47	4.49	1.43	1.51	2.48	1.18	1.40	2.10	0.86	0.75	1.31	0.75	0.57	0.95	0.50	0.43	0.90	0.28	0.29	0.54	0.22	0.32
39000000	6.21	6.22	11.16	4.99	6.47	8.42	4.68	3.54	6.48	3.60	3.08	4.72	2.26	2.47	4.49	1.43	1.51	2.49	1.18	1.40	2.10	0.87	0.75	1.31	0.75	0.57	0.95	0.50	0.43	0.90	0.28	0.29	0.54	0.22	0.32
40000000	6.21	6.21	11.16	4.99	6.47	8.41	4.68	3.54	6.48	3.60	3.08	4.72	2.26	2.47	4.49	1.43	1.51	2.49	1.18	1.40	2.10	0.87	0.75	1.31	0.75	0.57	0.96	0.50	0.43	0.90	0.28	0.29	0.54	0.22	0.32
41000000	6.21	6.21	11.15	4.99	6.47	8.41	4.68	3.54	6.48	3.60	3.08	4.72	2.26	2.47	4.49	1.43	1.51	2.49	1.18	1.40	2.10	0.87	0.75	1.31	0.75	0.57	0.96	0.50	0.43	0.90	0.29	0.29	0.54	0.22	0.32
42000000	6.20	6.20	11.14	4.98	6.46	8.41	4.68	3.54	6.48	3.60	3.08	4.72	2.26	2.47	4.50	1.43	1.51	2.49	1.18	1.40	2.11	0.87	0.75	1.31	0.75	0.57	0.96	0.50	0.43	0.90	0.29	0.29	0.54	0.22	0.32
43000000	6.20	6.20	11.14	4.98	6.46	8.41	4.67	3.54	6.48	3.60	3.08	4.72	2.27	2.47	4.50	1.43	1.51	2.49	1.18	1.40	2.11	0.87	0.75	1.31	0.75	0.57	0.96	0.50	0.43	0.90	0.29	0.29	0.54	0.22	0.32
44000000	6.20	6.20	11.13	4.98	6.46	8.40	4.67	3.54	6.48	3.60	3.08	4.72	2.27	2.47	4.50	1.43	1.51	2.49	1.18	1.41	2.11	0.87	0.76	1.31	0.75	0.57	0.96	0.50	0.43	0.90	0.29	0.29	0.54	0.22	0.32
45000000	6.19	6.19	11.12	4.98	6.46	8.40	4.67	3.53	6.48	3.60	3.08	4.72	2.27	2.47	4.50	1.44	1.51	2.49	1.18	1.41	2.11	0.87	0.76	1.32	0.75	0.57	0.96	0.50	0.43	0.90	0.29	0.29	0.54	0.22	0.32
46000000	6.19	6.19	11.12	4.97	6.45	8.40	4.67	3.53	6.47	3.60	3.08	4.72	2.27	2.47	4.50	1.44	1.51	2.49	1.18	1.41	2.11	0.87	0.76	1.32	0.76	0.57	0.96	0.50	0.43	0.91	0.29	0.29	0.54	0.22	0.32
47000000	6.19	6.19	11.11	4.97	6.45	8.39	4.67	3.53	6.47	3.60	3.08	4.72	2.27	2.47	4.50	1.44	1.51	2.49	1.18	1.41	2.11	0.87	0.76	1.32	0.76	0.57	0.96	0.50	0.44	0.91	0.29	0.29	0.55	0.22	0.32
48000000	6.18	6.18	11.11	4.97	6.45	8.39	4.67	3.53	6.47	3.60	3.08	4.72	2.27	2.47	4.50	1.44	1.51	2.50	1.18	1.41	2.11	0.87	0.76	1.32	0.76	0.57	0.96	0.50	0.44	0.91	0.29	0.29	0.55	0.23	0.32
49000000	6.18	6.18	11.10	4.97	6.45	8.39	4.67	3.53	6.47	3.60	3.08	4.72	2.27	2.47	4.50	1.44	1.51	2.50	1.18	1.41	2.12	0.87	0.76	1.32	0.76	0.57	0.97	0.50	0.44	0.91	0.29	0.29	0.55	0.23	0.32
50000000	6.18	6.18	11.10	4.97	6.44	8.39	4.67	3.53	6.47	3.59	3.08	4.72	2.27	2.47	4.50	1.44	1.51	2.50	1.19	1.41	2.12	0.87	0.76	1.32	0.76	0.57	0.97	0.51	0.44	0.91	0.29	0.29	0.55	0.23	0.32
51000000	6.17	6.17	11.09	4.96	6.44	8.38	4.66	3.53	6.47	3.59	3.08	4.72	2.27	2.47	4.51	1.44	1.51	2.50	1.19	1.41	2.12	0.87	0.76	1.32	0.76	0.58	0.97	0.51	0.44	0.91	0.29	0.30	0.55	0.23	0.32
52000000	6.17	6.17	11.09	4.96	6.44	8.38	4.66	3.53	6.47	3.59	3.08	4.72	2.27	2.47	4.51	1.44	1.51	2.50	1.19	1.41	2.12	0.88	0.76	1.32	0.76	0.58	0.97	0.51	0.44	0.91	0.29	0.30	0.55	0.23	0.32
53000000	6.17	6.17	11.08	4.96	6.44	8.38	4.66	3.53	6.47	3.59	3.08	4.72	2.27	2.47	4.51	1.44	1.52	2.50	1.19	1.41	2.12	0.88	0.76	1.32	0.76	0.58	0.97	0.51	0.44	0.91	0.29	0.30	0.55	0.23	0.32
54000000	6.16	6.16	11.08	4.96	6.43	8.38	4.66	3.53	6.47	3.59	3.08	4.72	2.27	2.47	4.51	1.44	1.52	2.50	1.19	1.41	2.12	0.88	0.76	1.33	0.76	0.58	0.97	0.51	0.44	0.92	0.29	0.30	0.55	0.23	0.33
55000000	6.16	6.16	11.07	4.96	6.43	8.37	4.66	3.53	6.47	3.59	3.08	4.72	2.27	2.47	4.51	1.44	1.52	2.50	1.19	1.41	2.12	0.88	0.76	1.33	0.76	0.58	0.97	0.51	0.44	0.92	0.29	0.30	0.55	0.23	0.33
56000000	6.16	6.16	11.07	4.95	6.43	8.37	4.66	3.53	6.47	3.59	3.08	4.72	2.27	2.47	4.51	1.44	1.52	2.51	1.19	1.42	2.12	0.88	0.76	1.33	0.76	0.58	0.97	0.51	0.44	0.92	0.29	0.30	0.55	0.23	0.33
57000000	6.15	6.15	11.06	4.95	6.43	8.37	4.66	3.52	6.46	3.59	3.08	4.72	2.27	2.47	4.51	1.44	1.52	2.51	1.19	1.42	2.13	0.88	0.76	1.33	0.76	0.58	0.97	0.51	0.44	0.92	0.29	0.30	0.55	0.23	0.33
58000000	6.15	6.15	11.06	4.95	6.42	8.37	4.66	3.52	6.46	3.59	3.08	4.72	2.27	2.47	4.51	1.44	1.52	2.51	1.19	1.42	2.13	0.88	0.76	1.33	0.76	0.58	0.98	0.51	0.44	0.92	0.29	0.30	0.55	0.23	0.33
59000000	6.15	6.15	11.05	4.95	6.42	8.36	4.66	3.52	6.46	3.59	3.08	4.72	2.27	2.47	4.51	1.44	1.52	2.51	1.19	1.42	2.13	0.88	0.76	1.33	0.76	0.58	0.98	0.51	0.44	0.92	0.29	0.30	0.56	0.23	0.33
60000000	6.14	6.15	11.05	4.95	6.42	8.36	4.65	3.52	6.46	3.59	3.08	4.72	2.27	2.47	4.51	1.44	1.52	2.51	1.19	1.42	2.13	0.88	0.77	1.33	0.77	0.58	0.98	0.51	0.44	0.92	0.29	0.30	0.56	0.23	0.33
61000000	6.14	6.14	11.04	4.95	6.42	8.36	4.65	3.52	6.46	3.59	3.08	4.72	2.27	2.47	4.52	1.44	1.52	2.51	1.19	1.42	2.13	0.88	0.77	1.33	0.77	0.58	0.98	0.51	0.44	0.92	0.29	0.30	0.56	0.23	0.33
62000000	6.14	6.14	11.04	4.94	6.41	8.36	4.65	3.52	6.46	3.59	3.08	4.72	2.27	2.47	4.52	1.44	1.52	2.51	1.19	1.42	2.13	0.88	0.77	1.33	0.77	0.58	0.98	0.51	0.44	0.92	0.29	0.30	0.56	0.23	0.33
63000000	6.14	6.14	11.03	4.94	6.41	8.36	4.65	3.52	6.46	3.59	3.08	4.72	2.27	2.48	4.52	1.45	1.52	2.51	1.19	1.42	2.13	0.88	0.77	1.33	0.77	0.58	0.98	0.51	0.44	0.92	0.29	0.30	0.56	0.23	0.33
64000000	6.13	6.13	11.03	4.94	6.41	8.35	4.65	3.52	6.46	3.59	3.08	4.72	2.27	2.48	4.52	1.45	1.52	2.51	1.19	1.42	2.13	0.88	0.77	1.34	0.77	0.58	0.98	0.51	0.44	0.93	0.30	0.30	0.56	0.23	0.33
65000000	6.13	6.13	11.02	4.94	6.41	8.35	4.65	3.52	6.46	3.59	3.08	4.72	2.27	2.48	4.52	1.45	1.52	2.51	1.19	1.42	2.14	0.88	0.77	1.34	0.77	0.58	0.98	0.51	0.44	0.93	0.30	0.30	0.56	0.23	0.33
66000000	6.13	6.13	11.02	4.94	6.40	8.35	4.65	3.52	6.46	3.59	3.08	4.72	2.27	2.48	4.52	1.45	1.52	2.52	1.19	1.42	2.14	0.88	0.77	1.34	0.77	0.58	0.98	0.51	0.44	0.93	0.30	0.30	0.56	0.23	0.33
67000000	6.12	6.13	11.02	4.94	6.40	8.35	4.65	3.52	6.46	3.59	3.08	4.72	2.27	2.48	4.52	1.45	1.52	2.52	1.19	1.42	2.14	0.88	0.77	1.34	0.77	0.58	0.98	0.51	0.44	0.93	0.30	0.30	0.56	0.23	0.33
68000000	6.12	6.12	11.01	4.93	6.40	8.35	4.64	3.52	6.46	3.59	3.08	4.72	2.27	2.48	4.52	1.45	1.52	2.52	1.19	1.42	2.14	0.88	0.77	1.34	0.77	0.59	0.98	0.51	0.44	0.93	0.30	0.30	0.56	0.23	0.33
69000000	6.12	6.12	11.01	4.93	6.40	8.34	4.64	3.52	6.45	3.59	3.08	4.72	2.27	2.48	4.52	1.45	1.52	2.52	1.20	1.42	2.14	0.88	0.77	1.34	0.77	0.59	0.98	0.51	0.45	0.93	0.30	0.30	0.56	0.23	0.33
70000000	6.12	6.12	11.00	4.93	6.39	8.34	4.64	3.52	6.45	3.59	3.08	4.72	2.27	2.48	4.52	1.45	1.52	2.52	1.20	1.43	2.14	0.89	0.77	1.34	0.77	0.59	0.99	0.52	0.45	0.93	0.30	0.30	0.56	0.23	0.33
71000000	6.11	6.12	11.00	4.93	6.39	8.34	4.64	3.51	6.45	3.59	3.08	4.72	2.27	2.48	4.52	1.45	1.52	2.52	1.20	1.43	2.14	0.89	0.77	1.34	0.77	0.59	0.99	0.52	0.45	0.93	0.30	0.30	0.56	0.23	0.33
72000000	6.11	6.11	11.00	4.93	6.39	8.34	4.64	3.51	6.45	3.59	3.08	4.72	2.27	2.48	4.53	1.45	1.52	2.52	1.20	1.43	2.14	0.89	0.77	1.34	0.77	0.59	0.99	0.52	0.45	0.93	0.30	0.30	0.56	0.23	0.33
73000000	6.11	6.11	10.99	4.93	6.39	8.34	4.64	3.51	6.45	3.59	3.08	4.72	2.27	2.48	4.53	1.45	1.52	2.52	1.20	1.43	2.15	0.89	0.77	1.34	0.77	0.59	0.99	0.52	0.45	0.93	0.30	0.30	0.57	0.23	0.33
74000000	6.11	6.11	10.99	4.92	6.39	8.33	4.64	3.51	6.45	3.59	3.08	4.72	2.27	2.48	4.53	1.45	1.53	2.52	1.20	1.43	2.15	0.89	0.77	1.35	0.77	0.59	0.99	0.52	0.45	0.93	0.30	0.30	0.57	0.23	0.33
75000000	6.10	6.11	10.98	4.92	6.38	8.33	4.64	3.51	6.45	3.59	3.08	4.72	2.27	2.48	4.53	1.45	1.53	2.52	1.20	1.43	2.15	0.89	0.77	1.35	0.77	0.59	0.99	0.52	0.45	0.94	0.30	0.30	0.57	0.23	0.33
76000000	6.10	6.10	10.98	4.92	6.38	8.33	4.64	3.51	6.45	3.59	3.08	4.72	2.27	2.48	4.53	1.45	1.53	2.52	1.20	1.43	2.15	0.89	0.77	1.35	0.77	0.59	0.99	0.52	0.45	0.94	0.30	0.30	0.57	0.23	0.34
77000000	6.10	6.10	10.97	4.92	6.38	8.33	4.64	3.51	6.45	3.59	3.08	4.72	2.27	2.48	4.53	1.45	1.53	2.53	1.20	1.43	2.15	0.89	0.77	1.35	0.78	0.59	0.99	0.52	0.45	0.94	0.30	0.30	0.57	0.23	0.34
78000000	6.10	6.10	10.97	4.92	6.38	8.33	4.63	3.51	6.45	3.59	3.07	4.72	2.27	2.48	4.53	1.45	1.53	2.53	1.20	1.43	2.15	0.89	0.77	1.35	0.78	0.59	0.99	0.52	0.45	0.94	0.30	0.30	0.57	0.23	0.34
79000000	6.09	6.10	10.97	4.92	6.38	8.32	4.63	3.51	6.45	3.59	3.07	4.72	2.27	2.48	4.53	1.45	1.53	2.53	1.20	1.43	2.15	0.89	0.77	1.35	0.78	0.59	0.99	0.52	0.45	0.94	0.30	0.30	0.57	0.23	0.34
80000000	6.09	6.09	10.96	4.92	6.37	8.32	4.63	3.51	6.45	3.59	3.07	4.72	2.27	2.48	4.53	1.45	1.53	2.53	1.20	1.43	2.15	0.89	0.77	1.35	0.78	0.59	0.99	0.52	0.45	0.94	0.30	0.31	0.57	0.23	0.34
81000000	6.09	6.09	10.96	4.91	6.37	8.32	4.63	3.51	6.45	3.59	3.07	4.72	2.27	2.48	4.53	1.45	1.53	2.53	1.20	1.43	2.15	0.89	0.78	1.35	0.78	0.59	0.99	0.52	0.45	0.94	0.30	0.31	0.57	0.23	0.34
82000000	6.09	6.09	10.96	4.91	6.37	8.32	4.63	3.51	6.45	3.59	3.07	4.72	2.27	2.48	4.53	1.45	1.53	2.53	1.20	1.43	2.16	0.89	0.78	1.35	0.78	0.59	1.00	0.52	0.45	0.94	0.30	0.31	0.57	0.24	0.34
83000000	6.08	6.09	10.95	4.91	6.37	8.32	4.63	3.51	6.45	3.59	3.07	4.72	2.27	2.48	4.53	1.45	1.53	2.53	1.20	1.43	2.16	0.89	0.78	1.35	0.78	0.59	1.00	0.52	0.45	0.94	0.30	0.31	0.57	0.24	0.34
84000000	6.08	6.08	10.95	4.91	6.37	8.31	4.63	3.51	6.45	3.59	3.07	4.72	2.27	2.48	4.54	1.45	1.53	2.53	1.20	1.43	2.16	0.89	0.78	1.35	0.78	0.59	1.00	0.52	0.45	0.94	0.30	0.31	0.57	0.24	0.34
85000000	6.08	6.08	10.95	4.91	6.37	8.31	4.63	3.51	6.44	3.59	3.07	4.72	2.27	2.48	4.54	1.45	1.53	2.53	1.20	1.43	2.16	0.89	0.78	1.35	0.78	0.59	1.00	0.52	0.45	0.94	0.30	0.31	0.57	0.24	0.34
86000000	6.08	6.08	10.94	4.91	6.36	8.31	4.63	3.51	6.44	3.59	3.07	4.72	2.27	2.48	4.54	1.45	1.53	2.53	1.20	1.43	2.16	0.89	0.78	1.35	0.78	0.59	1.00	0.52	0.45	0.95	0.30	0.31	0.57	0.24	0.34
87000000	6.07	6.08	10.94	4.91	6.36	8.31	4.63	3.51	6.44	3.59	3.07	4.72	2.27	2.48	4.54	1.45	1.53	2.53	1.20	1.43	2.16	0.89	0.78	1.36	0.78	0.59	1.00	0.52	0.45	0.95	0.30	0.31	0.57	0.24	0.34
88000000	6.07	6.07	10.93	4.90	6.36	8.31	4.62	3.51	6.44	3.59	3.07	4.73	2.28	2.48	4.54	1.45	1.53	2.53	1.20	1.44	2.16	0.89	0.78	1.36	0.78	0.59	1.00	0.52	0.45	0.95	0.30	0.31	0.57	0.24	0.34
89000000	6.07	6.07	10.93	4.90	6.36	8.31	4.62	3.50	6.44	3.59	3.07	4.73	2.28	2.48	4.54	1.45	1.53	2.54	1.20	1.44	2.16	0.89	0.78	1.36	0.78	0.59	1.00	0.52	0.45	0.95	0.30	0.31	0.57	0.24	0.34
90000000	6.07	6.07	10.93	4.90	6.36	8.30	4.62	3.50	6.44	3.59	3.07	4.73	2.28	2.48	4.54	1.46	1.53	2.54	1.20	1.44	2.16	0.89	0.78	1.36	0.78	0.59	1.00	0.52	0.45	0.95	0.30	0.31	0.57	0.24	0.34
91000000	6.06	6.07	10.92	4.90	6.35	8.30	4.62	3.50	6.44	3.59	3.07	4.73	2.28	2.48	4.54	1.46	1.53	2.54	1.20	1.44	2.16	0.89	0.78	1.36	0.78	0.60	1.00	0.52	0.45	0.95	0.30	0.31	0.57	0.24	0.34
92000000	6.06	6.06	10.92	4.90	6.35	8.30	4.62	3.50	6.44	3.59	3.07	4.73	2.28	2.48	4.54	1.46	1.53	2.54	1.21	1.44	2.16	0.90	0.78	1.36	0.78	0.60	1.00	0.52	0.45	0.95	0.30	0.31	0.58	0.24	0.34
93000000	6.06	6.06	10.92	4.90	6.35	8.30	4.62	3.50	6.44	3.59	3.07	4.73	2.28	2.48	4.54	1.46	1.53	2.54	1.21	1.44	2.17	0.90	0.78	1.36	0.78	0.60	1.00	0.53	0.45	0.95	0.30	0.31	0.58	0.24	0.34
94000000	6.06	6.06	10.91	4.90	6.35	8.30	4.62	3.50	6.44	3.59	3.07	4.73	2.28	2.48	4.55	1.46	1.53	2.54	1.21	1.44	2.17	0.90	0.78	1.36	0.78	0.60	1.00	0.53	0.45	0.95	0.30	0.31	0.58	0.24	0.34
95000000	6.06	6.06	10.91	4.89	6.35	8.30	4.62	3.50	6.44	3.59	3.07	4.73	2.28	2.48	4.55	1.46	1.54	2.54	1.21	1.44	2.17	0.90	0.78	1.36	0.78	0.60	1.00	0.53	0.45	0.95	0.30	0.31	0.58	0.24	0.34
96000000	6.05	6.06	10.91	4.89	6.35	8.29	4.62	3.50	6.44	3.59	3.07	4.73	2.28	2.48	4.55	1.46	1.54	2.54	1.21	1.44	2.17	0.90	0.78	1.36	0.78	0.60	1.01	0.53	0.45	0.95	0.30	0.31	0.58	0.24	0.34
97000000	6.05	6.05	10.90	4.89	6.34	8.29	4.62	3.50	6.44	3.59	3.07	4.73	2.28	2.48	4.55	1.46	1.54	2.54	1.21	1.44	2.17	0.90	0.78	1.36	0.79	0.60	1.01	0.53	0.45	0.95	0.31	0.31	0.58	0.24	0.34
98000000	6.05	6.05	10.90	4.89	6.34	8.29	4.62	3.50	6.43	3.59	3.07	4.73	2.28	2.48	4.55	1.46	1.54	2.54	1.21	1.44	2.17	0.90	0.78	1.36	0.79	0.60	1.01	0.53	0.46	0.96	0.31	0.31	0.58	0.24	0.34
99000000	6.05	6.05	10.90	4.89	6.34	8.29	4.62	3.50	6.43	3.59	3.07	4.73	2.28	2.48	4.55	1.46	1.54	2.54	1.21	1.44	2.17	0.90	0.78	1.36	0.79	0.60	1.01	0.53	0.46	0.96	0.31	0.31	0.58	0.24	0.34
100000000	6.05	6.05	10.90	4.89	6.34	8.29	4.61	3.50	6.43	3.59	3.07	4.73	2.28	2.48	4.55	1.46	1.54	2.54	1.21	1.44	2.17	0.90	0.78	1.36	0.79	0.60	1.01	0.53	0.46	0.96	0.31	0.31	0.58	0.24	0.35
101000000	6.04	6.05	10.89	4.89	6.34	8.29	4.61	3.50	6.43	3.59	3.07	4.73	2.28	2.48	4.55	1.46	1.54	2.54	1.21	1.44	2.17	0.90	0.78	1.37	0.79	0.60	1.01	0.53	0.46	0.96	0.31	0.31	0.58	0.24	0.35
102000000	6.04	6.04	10.89	4.89	6.34	8.28	4.61	3.50	6.43	3.59	3.07	4.73	2.28	2.48	4.55	1.46	1.54	2.54	1.21	1.44	2.17	0.90	0.78	1.37	0.79	0.60	1.01	0.53	0.46	0.96	0.31	0.31	0.58	0.24	0.35
103000000	0.04	0.04	10.89	4.88	0.33	8.28	4.01	3.50	0.43	3.59	3.07	4.73	2.28	2.48	4.55	1.40	1.54	2.55	1.21	1.44	2.17	0.90	0.78	1.37	0.79	0.00	1.01	0.53	0.40	0.90	0.31	0.31	0.58	0.24	0.35

Figure 3.
Gaps frequency in intervals of 100 million primes, step 1 million.

one such sequence, with its starting point indicated in the first column. The first row indicates the gap code corresponding to each column.

Figure 4 presents a similar view, showing the normalized frequencies for the first 42 gaps in the second scenario: sequences of 150 million consecutive primes. As one can observe, the distribution patterns are essentially identical.

Here are several interesting, apparent findings:

- the pattern of the gaps frequencies is relatively consistent for each large enough interval from which the gaps are sampled—each gap has practically the same distribution over sufficiently large intervals that are close enough, regardless of the interval;
- using a logarithmic base, the gaps align almost linearly, with a slowly diminishing slope as the sets shift toward higher absolute values;
- each third gap is significantly more frequent than the preceding two; and
- the frequency of gaps less than 20 (gaps 1 through 10) decreases slowly as the starting values of the prime intervals increase, while the frequency of other gaps increases very slowly.

Judging from these values, the frequency of gap 1 (the gap corresponding with twin primes) in any interval large enough is more or less constant—it decreases gently and monotonically as the values increase, but still remains one of the most frequent gaps. As observed in the tables, the most frequent gap codes are 3, 6, 5, and 9, followed by codes 1 (corresponding to twin primes) and 2 (the actual prime gaps are twice these values).

2.1 Conjectures on prime gap distributions

2.1.1 Stability across intervals

Let $G_n = \left\{g_i^{(n)}\right\}_{i=1}^{k}$ denote the normalized frequency distribution of gap codes g_i in a sequence of n consecutive primes, sampled from an interval I_n. Then, for sufficiently large n and for intervals I_n and I_{n+1} such that $|I_n \cap I_{n+1}|$ is non-negligible, we have:

$$G_n \approx G_{n+1} \tag{1}$$

in the sense of distributional similarity (e.g., small total variation distance). That is, the distribution of prime gaps remains stable across large, closely spaced intervals.

2.1.2 Logarithmic alignment

Let g_i be the frequency of the i-th gap code over a fixed interval of n consecutive primes. Then, plotting $\log(g_i)$ versus i produces a curve that approximates a decreasing affine function for low-to-mid values of i, that is

$$\log(g_i) \approx a - bi, \quad \text{with } b > 0 \tag{2}$$

The slope b decreases slowly as the starting point of the interval increases.

Start	1	2	3	4	5	6	7	8	9	10	11	12	13	14	15	16	17	18	19	20	21	22	23	24	25	26	27	28	29	30	31	32	33	34	35	36	37	38	39	40	41	42
0	6.36	6.36	11.39	5.08	6.58	8.52	4.73	3.56	6.51	3.59	3.08	4.70	2.25	2.45	4.43	1.41	1.48	2.43	1.15	1.37	2.04	0.84	0.73	1.26	0.72	0.54	0.91	0.48	0.41	0.85	0.27	0.27	0.51	0.21	0.30	0.30	0.15	0.13	0.26	0.13	0.09	0.20
1,000,000	6.35	6.35	11.37	5.07	6.57	8.51	4.72	3.56	6.51	3.59	3.08	4.70	2.25	2.45	4.44	1.41	1.48	2.44	1.15	1.37	2.04	0.84	0.73	1.26	0.72	0.55	0.92	0.48	0.41	0.85	0.27	0.28	0.51	0.21	0.30	0.31	0.15	0.14	0.26	0.13	0.10	0.20
2,000,000	6.34	6.34	11.35	5.06	6.56	8.50	4.72	3.56	6.51	3.59	3.08	4.70	2.25	2.45	4.44	1.41	1.49	2.44	1.16	1.37	2.05	0.84	0.73	1.27	0.73	0.55	0.92	0.48	0.41	0.86	0.27	0.28	0.51	0.21	0.30	0.31	0.15	0.14	0.26	0.13	0.10	0.20
3,000,000	6.33	6.33	11.33	5.06	6.55	8.50	4.72	3.56	6.51	3.59	3.08	4.70	2.25	2.45	4.45	1.41	1.49	2.45	1.16	1.37	2.05	0.84	0.73	1.27	0.73	0.55	0.92	0.48	0.42	0.86	0.27	0.28	0.51	0.21	0.30	0.31	0.15	0.14	0.26	0.14	0.10	0.20
4,000,000	6.32	6.32	11.32	5.05	6.55	8.49	4.71	3.56	6.50	3.59	3.08	4.71	2.25	2.45	4.45	1.42	1.49	2.45	1.16	1.38	2.06	0.85	0.73	1.27	0.73	0.55	0.93	0.48	0.42	0.86	0.27	0.28	0.52	0.21	0.30	0.31	0.15	0.14	0.26	0.14	0.10	0.20
5,000,000	6.31	6.31	11.31	5.05	6.54	8.49	4.71	3.56	6.50	3.59	3.08	4.71	2.25	2.46	4.45	1.42	1.49	2.45	1.16	1.38	2.06	0.85	0.73	1.28	0.73	0.55	0.93	0.48	0.42	0.87	0.27	0.28	0.52	0.21	0.30	0.31	0.15	0.14	0.26	0.14	0.10	0.20
6,000,000	6.30	6.30	11.30	5.04	6.54	8.48	4.71	3.55	6.50	3.59	3.08	4.71	2.25	2.46	4.46	1.42	1.49	2.46	1.16	1.38	2.06	0.85	0.74	1.28	0.73	0.55	0.93	0.48	0.42	0.87	0.27	0.28	0.52	0.21	0.31	0.31	0.15	0.14	0.27	0.14	0.10	0.20
7,000,000	6.29	6.29	11.28	5.04	6.53	8.47	4.71	3.55	6.50	3.60	3.08	4.71	2.25	2.46	4.46	1.42	1.49	2.46	1.16	1.38	2.07	0.85	0.74	1.28	0.73	0.55	0.93	0.49	0.42	0.87	0.28	0.28	0.52	0.21	0.31	0.31	0.16	0.14	0.27	0.14	0.10	0.20
8,000,000	6.29	6.29	11.27	5.03	6.53	8.47	4.70	3.55	6.50	3.60	3.08	4.71	2.25	2.46	4.46	1.42	1.49	2.46	1.16	1.38	2.07	0.85	0.74	1.28	0.74	0.56	0.93	0.49	0.42	0.87	0.28	0.28	0.52	0.21	0.31	0.31	0.16	0.14	0.27	0.14	0.10	0.20
9,000,000	6.28	6.28	11.26	5.03	6.52	8.46	4.70	3.55	6.50	3.60	3.08	4.71	2.26	2.46	4.46	1.42	1.50	2.46	1.17	1.39	2.07	0.85	0.74	1.29	0.74	0.56	0.94	0.49	0.42	0.88	0.28	0.28	0.52	0.22	0.31	0.32	0.16	0.14	0.27	0.14	0.10	0.21
10,000,000	6.27	6.27	11.25	5.03	6.52	8.46	4.70	3.55	6.49	3.60	3.08	4.71	2.26	2.46	4.47	1.42	1.50	2.46	1.17	1.39	2.07	0.85	0.74	1.29	0.74	0.56	0.94	0.49	0.42	0.88	0.28	0.28	0.53	0.22	0.31	0.32	0.16	0.14	0.27	0.14	0.10	0.21
11,000,000	6.27	6.27	11.24	5.02	6.51	8.46	4.70	3.55	6.49	3.59	3.08	4.71	2.26	2.46	4.47	1.42	1.50	2.47	1.17	1.39	2.08	0.86	0.74	1.29	0.74	0.56	0.94	0.49	0.42	0.88	0.28	0.28	0.53	0.22	0.31	0.32	0.16	0.14	0.27	0.14	0.10	0.21
12,000,000	6.26	6.26	11.23	5.02	6.51	8.45	4.69	3.55	6.49	3.59	3.08	4.71	2.26	2.46	4.47	1.42	1.50	2.47	1.17	1.39	2.08	0.86	0.74	1.29	0.74	0.56	0.94	0.49	0.42	0.88	0.28	0.28	0.53	0.22	0.31	0.32	0.16	0.14	0.27	0.14	0.10	0.21
13,000,000	6.26	6.26	11.23	5.02	6.50	8.45	4.69	3.55	6.49	3.59	3.08	4.71	2.26	2.46	4.47	1.43	1.50	2.47	1.17	1.39	2.08	0.86	0.74	1.29	0.74	0.56	0.94	0.49	0.42	0.88	0.28	0.29	0.53	0.22	0.31	0.32	0.16	0.14	0.27	0.14	0.10	0.21
14,000,000	6.25	6.25	11.22	5.01	6.50	8.44	4.69	3.54	6.49	3.59	3.08	4.71	2.26	2.46	4.48	1.43	1.50	2.47	1.17	1.39	2.08	0.86	0.75	1.29	0.74	0.56	0.94	0.49	0.43	0.89	0.28	0.29	0.53	0.22	0.31	0.32	0.16	0.14	0.27	0.14	0.10	0.21
15,000,000	6.25	6.25	11.21	5.01	6.50	8.44	4.69	3.54	6.49	3.59	3.08	4.71	2.26	2.46	4.48	1.43	1.50	2.47	1.17	1.39	2.09	0.86	0.75	1.30	0.74	0.56	0.95	0.49	0.43	0.89	0.28	0.29	0.53	0.22	0.31	0.32	0.16	0.14	0.27	0.14	0.10	0.21
16,000,000	6.24	6.24	11.20	5.01	6.49	8.44	4.69	3.54	6.49	3.59	3.08	4.71	2.26	2.46	4.48	1.43	1.50	2.48	1.17	1.40	2.09	0.86	0.75	1.30	0.74	0.56	0.95	0.49	0.43	0.89	0.28	0.29	0.53	0.22	0.31	0.32	0.16	0.14	0.27	0.14	0.10	0.21
17,000,000	6.24	6.24	11.19	5.00	6.49	8.43	4.69	3.54	6.49	3.59	3.08	4.71	2.26	2.46	4.48	1.43	1.50	2.48	1.17	1.40	2.09	0.86	0.75	1.30	0.75	0.56	0.95	0.50	0.43	0.89	0.28	0.29	0.53	0.22	0.31	0.32	0.16	0.14	0.28	0.14	0.10	0.21
18,000,000	6.23	6.23	11.19	5.00	6.48	8.43	4.68	3.54	6.48	3.59	3.08	4.72	2.26	2.46	4.48	1.43	1.50	2.48	1.17	1.40	2.09	0.86	0.75	1.30	0.75	0.57	0.95	0.50	0.43	0.89	0.28	0.29	0.54	0.22	0.32	0.32	0.16	0.14	0.28	0.14	0.10	0.21
19,000,000	6.23	6.23	11.18	5.00	6.48	8.42	4.68	3.54	6.48	3.59	3.08	4.72	2.26	2.46	4.49	1.43	1.51	2.48	1.18	1.40	2.09	0.86	0.75	1.30	0.75	0.57	0.95	0.50	0.43	0.89	0.28	0.29	0.54	0.22	0.32	0.32	0.16	0.14	0.28	0.14	0.10	0.21
20,000,000	6.22	6.22	11.17	4.99	6.48	8.42	4.68	3.54	6.48	3.59	3.08	4.72	2.26	2.47	4.49	1.43	1.51	2.48	1.18	1.40	2.10	0.86	0.75	1.31	0.75	0.57	0.95	0.50	0.43	0.90	0.28	0.29	0.54	0.22	0.32	0.33	0.16	0.14	0.28	0.14	0.10	0.21
21,000,000	6.22	6.22	11.16	4.99	6.47	8.42	4.68	3.54	6.48	3.59	3.08	4.72	2.26	2.47	4.49	1.43	1.51	2.48	1.18	1.40	2.10	0.87	0.75	1.31	0.75	0.57	0.96	0.50	0.43	0.90	0.28	0.29	0.54	0.22	0.32	0.33	0.16	0.14	0.28	0.14	0.10	0.21
22,000,000	6.21	6.21	11.16	4.99	6.47	8.41	4.68	3.54	6.48	3.59	3.08	4.72	2.26	2.47	4.49	1.43	1.51	2.49	1.18	1.40	2.10	0.87	0.75	1.31	0.75	0.57	0.96	0.50	0.43	0.90	0.29	0.29	0.54	0.22	0.32	0.33	0.16	0.15	0.28	0.14	0.10	0.21
23,000,000	6.21	6.21	11.15	4.99	6.47	8.41	4.68	3.54	6.48	3.59	3.08	4.72	2.26	2.47	4.49	1.43	1.51	2.49	1.18	1.40	2.10	0.87	0.75	1.31	0.75	0.57	0.96	0.50	0.43	0.90	0.29	0.29	0.54	0.22	0.32	0.33	0.16	0.15	0.28	0.14	0.10	0.21
24,000,000	6.20	6.20	11.14	4.98	6.46	8.41	4.67	3.53	6.48	3.59	3.08	4.72	2.26	2.47	4.49	1.43	1.51	2.49	1.18	1.40	2.10	0.87	0.75	1.31	0.75	0.57	0.96	0.50	0.43	0.90	0.29	0.29	0.54	0.22	0.32	0.33	0.16	0.15	0.28	0.14	0.10	0.22
25,000,000	6.20	6.20	11.14	4.98	6.46	8.41	4.67	3.53	6.48	3.59	3.08	4.72	2.26	2.47	4.49	1.43	1.51	2.49	1.18	1.40	2.11	0.87	0.75	1.31	0.75	0.57	0.96	0.50	0.43	0.90	0.29	0.29	0.54	0.22	0.32	0.33	0.16	0.15	0.28	0.15	0.10	0.22
26,000,000	6.20	6.20	11.13	4.98	6.46	8.40	4.67	3.53	6.48	3.59	3.08	4.72	2.26	2.47	4.50	1.44	1.51	2.49	1.18	1.41	2.11	0.87	0.76	1.31	0.75	0.57	0.96	0.50	0.43	0.90	0.29	0.29	0.54	0.22	0.32	0.33	0.16	0.15	0.28	0.15	0.10	0.22
27,000,000	6.19	6.19	11.12	4.98	6.45	8.40	4.67	3.53	6.47	3.59	3.08	4.72	2.27	2.47	4.50	1.44	1.51	2.49	1.18	1.41	2.11	0.87	0.76	1.32	0.76	0.57	0.96	0.50	0.43	0.91	0.29	0.29	0.54	0.22	0.32	0.33	0.16	0.15	0.28	0.15	0.10	0.22
28,000,000	6.19	6.19	11.12	4.97	6.45	8.40	4.67	3.53	6.47	3.59	3.08	4.72	2.27	2.47	4.50	1.44	1.51	2.49	1.18	1.41	2.11	0.87	0.76	1.32	0.76	0.57	0.96	0.50	0.43	0.91	0.29	0.29	0.55	0.22	0.32	0.33	0.16	0.15	0.28	0.15	0.10	0.22
29,000,000	6.18	6.18	11.11	4.97	6.45	8.39	4.67	3.53	6.47	3.59	3.08	4.72	2.27	2.47	4.50	1.44	1.51	2.50	1.18	1.41	2.11	0.87	0.76	1.32	0.76	0.57	0.97	0.50	0.44	0.91	0.29	0.29	0.55	0.22	0.32	0.33	0.16	0.15	0.28	0.15	0.10	0.22
30,000,000	6.18	6.18	11.11	4.97	6.45	8.39	4.67	3.53	6.47	3.59	3.08	4.72	2.27	2.47	4.50	1.44	1.51	2.50	1.18	1.41	2.11	0.87	0.76	1.32	0.76	0.57	0.97	0.50	0.44	0.91	0.29	0.29	0.55	0.23	0.32	0.33	0.16	0.15	0.28	0.15	0.10	0.22
31,000,000	6.18	6.18	11.10	4.97	6.44	8.39	4.66	3.53	6.47	3.59	3.08	4.72	2.27	2.47	4.50	1.44	1.51	2.50	1.18	1.41	2.12	0.87	0.76	1.32	0.76	0.58	0.97	0.51	0.44	0.91	0.29	0.30	0.55	0.23	0.32	0.33	0.16	0.15	0.28	0.15	0.10	0.22
32,000,000	6.17	6.17	11.09	4.97	6.44	8.38	4.66	3.53	6.47	3.59	3.08	4.72	2.27	2.47	4.50	1.44	1.51	2.50	1.19	1.41	2.12	0.87	0.76	1.32	0.76	0.58	0.97	0.51	0.44	0.91	0.29	0.30	0.55	0.23	0.32	0.33	0.17	0.15	0.28	0.15	0.10	0.22
33,000,000	6.17	6.17	11.09	4.96	6.44	8.38	4.66	3.53	6.47	3.59	3.08	4.72	2.27	2.47	4.51	1.44	1.52	2.50	1.19	1.41	2.12	0.87	0.76	1.32	0.76	0.58	0.97	0.51	0.44	0.91	0.29	0.30	0.55	0.23	0.32	0.33	0.17	0.15	0.29	0.15	0.10	0.22
34,000,000	6.17	6.17	11.08	4.96	6.43	8.38	4.66	3.53	6.47	3.59	3.08	4.72	2.27	2.47	4.51	1.44	1.52	2.50	1.19	1.41	2.12	0.88	0.76	1.32	0.76	0.58	0.97	0.51	0.44	0.91	0.29	0.30	0.55	0.23	0.33	0.33	0.17	0.15	0.29	0.15	0.11	0.22
35,000,000	6.16	6.16	11.08	4.96	6.43	8.38	4.66	3.53	6.47	3.59	3.08	4.72	2.27	2.47	4.51	1.44	1.52	2.50	1.19	1.41	2.12	0.88	0.76	1.33	0.76	0.58	0.97	0.51	0.44	0.92	0.29	0.30	0.55	0.23	0.33	0.34	0.17	0.15	0.29	0.15	0.11	0.22
36,000,000	6.16	6.16	11.07	4.96	6.43	8.37	4.66	3.53	6.47	3.59	3.08	4.72	2.27	2.47	4.51	1.44	1.52	2.50	1.19	1.41	2.12	0.88	0.76	1.33	0.76	0.58	0.97	0.51	0.44	0.92	0.29	0.30	0.55	0.23	0.33	0.34	0.17	0.15	0.29	0.15	0.11	0.22
37,000,000	6.15	6.16	11.07	4.95	6.43	8.37	4.66	3.52	6.46	3.59	3.08	4.72	2.27	2.47	4.51	1.44	1.52	2.51	1.19	1.42	2.12	0.88	0.76	1.33	0.76	0.58	0.97	0.51	0.44	0.92	0.29	0.30	0.55	0.23	0.33	0.34	0.17	0.15	0.29	0.15	0.11	0.22
38,000,000	6.15	6.15	11.06	4.95	6.42	8.37	4.66	3.52	6.46	3.59	3.08	4.72	2.27	2.47	4.51	1.44	1.52	2.51	1.19	1.42	2.13	0.88	0.76	1.33	0.76	0.58	0.97	0.51	0.44	0.92	0.29	0.30	0.55	0.23	0.33	0.34	0.17	0.15	0.29	0.15	0.11	0.22
39,000,000	6.15	6.15	11.06	4.95	6.42	8.37	4.65	3.52	6.46	3.59	3.08	4.72	2.27	2.47	4.51	1.44	1.52	2.51	1.19	1.42	2.13	0.88	0.76	1.33	0.76	0.58	0.98	0.51	0.44	0.92	0.29	0.30	0.55	0.23	0.33	0.34	0.17	0.15	0.29	0.15	0.11	0.22
40,000,000	6.15	6.15	11.05	4.95	6.42	8.36	4.65	3.52	6.46	3.59	3.08	4.72	2.27	2.47	4.51	1.44	1.52	2.51	1.19	1.42	2.13	0.88	0.76	1.33	0.76	0.58	0.98	0.51	0.44	0.92	0.29	0.30	0.56	0.23	0.33	0.34	0.17	0.15	0.29	0.15	0.11	0.22
41,000,000	6.14	6.14	11.05	4.95	6.42	8.36	4.65	3.52	6.46	3.59	3.08	4.72	2.27	2.47	4.52	1.44	1.52	2.51	1.19	1.42	2.13	0.88	0.77	1.33	0.77	0.58	0.98	0.51	0.44	0.92	0.29	0.30	0.56	0.23	0.33	0.34	0.17	0.15	0.29	0.15	0.11	0.22
42,000,000	6.14	6.14	11.04	4.94	6.41	8.36	4.65	3.52	6.46	3.59	3.08	4.72	2.27	2.47	4.52	1.44	1.52	2.51	1.19	1.42	2.13	0.88	0.77	1.33	0.77	0.58	0.98	0.51	0.44	0.92	0.29	0.30	0.56	0.23	0.33	0.34	0.17	0.15	0.29	0.15	0.11	0.22
43,000,000	6.14	6.14	11.04	4.94	6.41	8.36	4.65	3.52	6.46	3.59	3.08	4.72	2.27	2.47	4.52	1.44	1.52	2.51	1.19	1.42	2.13	0.88	0.77	1.33	0.77	0.58	0.98	0.51	0.44	0.92	0.29	0.30	0.56	0.23	0.33	0.34	0.17	0.15	0.29	0.15	0.11	0.22
44,000,000	6.13	6.13	11.03	4.94	6.41	8.36	4.65	3.52	6.46	3.59	3.08	4.72	2.27	2.48	4.52	1.45	1.52	2.51	1.19	1.42	2.13	0.88	0.77	1.34	0.77	0.58	0.98	0.51	0.44	0.93	0.29	0.30	0.56	0.23	0.33	0.34	0.17	0.15	0.29	0.15	0.11	0.23
45,000,000	6.13	6.13	11.03	4.94	6.41	8.35	4.65	3.52	6.46	3.59	3.08	4.72	2.27	2.48	4.52	1.45	1.52	2.51	1.19	1.42	2.14	0.88	0.77	1.34	0.77	0.58	0.98	0.51	0.44	0.93	0.30	0.30	0.56	0.23	0.33	0.34	0.17	0.15	0.29	0.15	0.11	0.23
46,000,000	6.13	6.13	11.02	4.94	6.40	8.35	4.65	3.52	6.46	3.59	3.08	4.72	2.27	2.48	4.52	1.45	1.52	2.51	1.19	1.42	2.14	0.88	0.77	1.34	0.77	0.58	0.98	0.51	0.44	0.93	0.30	0.30	0.56	0.23	0.33	0.34	0.17	0.15	0.29	0.15	0.11	0.23
47,000,000	6.12	6.12	11.02	4.93	6.40	8.35	4.65	3.52	6.46	3.59	3.08	4.72	2.27	2.48	4.52	1.45	1.52	2.52	1.19	1.42	2.14	0.88	0.77	1.34	0.77	0.58	0.98	0.51	0.44	0.93	0.30	0.30	0.56	0.23	0.33	0.34	0.17	0.15	0.29	0.15	0.11	0.23
48,000,000	6.12	6.12	11.01	4.93	6.40	8.35	4.64	3.52	6.46	3.59	3.08	4.72	2.27	2.48	4.52	1.45	1.52	2.52	1.19	1.42	2.14	0.88	0.77	1.34	0.77	0.58	0.98	0.51	0.44	0.93	0.30	0.30	0.56	0.23	0.33	0.34	0.17	0.15	0.29	0.15	0.11	0.23
49,000,000	6.12	6.12	11.01	4.93	6.40	8.34	4.64	3.52	6.45	3.59	3.08	4.72	2.27	2.48	4.52	1.45	1.52	2.52	1.20	1.42	2.14	0.88	0.77	1.34	0.77	0.59	0.99	0.52	0.45	0.93	0.30	0.30	0.56	0.23	0.33	0.34	0.17	0.15	0.29	0.15	0.11	0.23
50,000,000	6.12	6.12	11.00	4.93	6.40	8.34	4.64	3.52	6.45	3.59	3.08	4.72	2.27	2.48	4.53	1.45	1.52	2.52	1.20	1.42	2.14	0.88	0.77	1.34	0.77	0.59	0.99	0.52	0.45	0.93	0.30	0.30	0.56	0.23	0.33	0.34	0.17	0.15	0.29	0.15	0.11	0.23
51,000,000	6.11	6.11	11.00	4.93	6.39	8.34	4.64	3.51	6.45	3.59	3.08	4.72	2.27	2.48	4.53	1.45	1.52	2.52	1.20	1.43	2.14	0.89	0.77	1.34	0.77	0.59	0.99	0.52	0.45	0.93	0.30	0.30	0.56	0.23	0.33	0.34	0.17	0.15	0.29	0.15	0.11	0.23
52,000,000	6.11	6.11	11.00	4.93	6.39	8.34	4.64	3.51	6.45	3.59	3.08	4.72	2.27	2.48	4.53	1.45	1.53	2.52	1.20	1.43	2.14	0.89	0.77	1.34	0.77	0.59	0.99	0.52	0.45	0.93	0.30	0.30	0.56	0.23	0.33	0.34	0.17	0.15	0.29	0.15	0.11	0.23
53,000,000	6.11	6.11	10.99	4.92	6.39	8.34	4.64	3.51	6.45	3.59	3.08	4.72	2.27	2.48	4.53	1.45	1.53	2.52	1.20	1.43	2.15	0.89	0.77	1.34	0.77	0.59	0.99	0.52	0.45	0.93	0.30	0.30	0.56	0.23	0.33	0.34	0.17	0.15	0.30	0.15	0.11	0.23

Figure 4.
Gaps frequency in intervals of 150 million primes, step 1 million.

2.1.3 Triplet frequency bias

Let g_i denote the frequency of gap code i. Then, for all $i \in \mathbb{N}$ such that $i \equiv 0 \pmod{3}$, we have:

$$g_i > \max(g_{i-1}, g_{i-2}) \tag{3}$$

That is, every third gap code (multiples of 3) occurs significantly more frequently than the two preceding codes.

2.1.4 Frequency shift of small gaps

Let $G_n^{(s)}$ be the normalized frequency of small gap codes $s \in \{1, 2, \ldots, 10\}$ in a sample of n consecutive primes. Then:

$$\frac{d}{dx} G_n^{(s)}(x) < 0 \quad \text{and} \quad \frac{d}{dx} G_n^{(l)}(x) > 0 \tag{4}$$

where x denotes the starting point of the interval and $G_n^{(l)}$ the frequency of larger gap codes. In other words, the frequency of small gaps decreases slowly as we move to higher intervals, while the frequency of larger gaps increases slowly.

3. Tandem gaps

Once we have a fast prime sequence generator, we can explore these sequences to identify new and interesting patterns, which can be validated across multiple numerical ranges. For instance, in an effort to extract higher values from second-order randomness, we analyzed the frequency of consecutive gap pairs—using the product of two successive gaps as the random value. We refer to such an ordered pair as a *tandem gap*. Unlike single gaps, whose values are bounded to 256 codes, tandem gaps can theoretically reach values up to 256^2 when multiplying the two constitutive gaps.

3.1 Impossible pairs

Upon analyzing the results, we observed a peculiar phenomenon: Beneath the seemingly random distribution lies a surprisingly uniform pattern, consistently preserved across all generated samples, as illustrated in **Figure 5**. The figure shows a matrix of absolute tandem gap frequencies, covering the first 30 × 50 combinations—with the first gap represented by columns and the second by rows—computed from the initial sequence (starting at zero) of 150 million primes in our second scenario.

We expected that all possible pairs will exist in our distribution—instead, we definitely see some rules emerging:

- some pairs are impossible (the frequency of that pair will be zero, the pair does not occur);
- the matrix shows that if pair (a, b) is impossible, pair (b, a) is also impossible;
- all pairs (a, a) are impossible except when a is a multiple of 3; and

	1	2	3	4	5	6	7	8	9	10	11	12	13	14	15	16	17	18	19	20	21	22	23	24	25	26	27	28	29	30
1	-	998,611	929,541	-	1,225,635	898,289	-	608,517	785,161	-	521,227	454,501	-	631,586	418,143	-	255,469	230,482	-	324,680	202,186	-	128,390	157,399	-	92,165	102,241	-	103,052	83,882
2	998,949	-	1,358,779	796,491	-	679,842	828,711	-	626,228	681,876	-	506,375	558,990	-	448,328	241,076	-	310,959	229,550	-	184,932	144,771	-	115,214	141,435	-	86,591	126,710	-	84,009
3	928,929	1,357,948	1,591,186	979,028	1,047,875	1,270,908	1,045,427	590,723	937,278	654,048	591,012	997,909	317,229	381,327	735,755	233,464	334,953	426,624	195,135	227,008	324,124	189,192	115,926	192,114	152,660	89,795	200,831	70,802	62,868	160,819
4	-	795,154	979,727	-	959,008	787,359	-	464,366	471,219	-	590,236	290,513	-	318,424	355,146	-	232,373	181,799	-	197,561	206,083	-	99,995	100,982	-	108,604	56,045	-	57,689	66,757
5	1,226,753	-	1,046,177	958,421	-	943,635	895,901	-	759,151	663,741	-	449,267	420,571	-	422,616	334,931	-	230,712	216,246	-	219,329	157,763	-	128,478	134,640	-	87,543	103,383	-	97,730
6	898,273	680,532	1,267,160	786,102	944,812	915,550	577,216	520,019	1,017,751	431,884	356,909	499,214	266,725	405,704	661,842	164,046	159,800	261,853	183,401	173,608	227,391	115,324	90,394	199,860	86,392	60,155	108,089	54,368	80,577	98,538
7	-	828,416	1,045,061	-	894,537	577,509	-	642,883	366,430	-	366,422	312,810	-	282,241	333,883	-	185,967	218,498	-	182,610	126,073	-	137,753	73,441	-	76,063	59,555	-	45,515	60,868
8	609,721	-	590,427	464,570	-	519,106	641,999	-	301,029	371,915	-	313,456	239,625	-	228,440	127,161	-	110,363	106,551	-	109,364	113,067	-	60,017	82,632	-	74,590	39,437	-	43,790
9	786,052	627,541	939,360	471,518	757,781	1,015,563	365,991	301,443	525,412	364,996	371,508	449,060	209,371	202,514	395,257	172,249	127,364	201,318	119,715	129,482	256,062	63,409	60,592	113,032	68,464	81,516	66,258	42,612	31,378	83,746
10	-	681,543	653,129	-	664,827	431,809	-	372,667	365,696	-	381,397	263,918	-	250,052	221,298	-	155,353	155,028	-	135,736	108,221	-	84,383	66,191	-	51,264	48,590	-	44,201	43,493
11	521,865	-	591,516	588,798	-	357,309	366,193	-	372,477	380,509	-	168,717	173,674	-	201,083	106,831	-	100,362	127,597	-	79,950	62,322	-	88,906	57,678	-	33,400	40,552	-	35,264
12	455,120	506,697	999,920	290,957	449,874	498,352	312,235	312,863	448,429	264,417	168,723	271,918	198,802	150,775	289,824	108,333	97,334	206,825	64,763	90,393	136,481	54,154	77,173	66,690	49,976	29,987	59,658	41,214	22,259	67,282
13	-	560,560	316,782	-	419,855	266,702	-	240,129	209,644	-	174,189	198,069	-	145,333	175,131	-	123,817	62,276	-	93,229	60,077	-	36,944	36,938	-	30,820	37,164	-	28,056	26,323
14	631,564	-	381,628	319,115	-	404,941	283,494	-	201,108	249,261	-	151,698	145,433	-	164,325	129,482	-	75,792	77,682	-	93,157	49,649	-	38,812	53,253	-	26,936	28,807	-	31,451
15	416,925	448,289	735,957	355,740	424,362	661,053	334,854	228,368	394,278	221,584	200,649	290,682	174,533	164,757	268,349	85,844	97,079	166,395	73,318	102,353	125,599	53,516	46,348	83,783	43,271	34,410	67,003	29,719	25,305	48,977
16	-	241,341	233,440	-	335,245	163,504	-	127,003	172,473	-	106,908	108,655	-	129,300	85,769	-	57,675	43,842	-	50,956	39,156	-	28,899	32,287	-	23,896	15,741	-	19,796	17,404
17	255,348	-	334,942	232,004	-	160,560	185,461	-	127,753	155,736	-	96,573	124,225	-	97,101	58,496	-	81,492	39,747	-	38,721	36,223	-	23,188	36,228	-	17,626	25,006	-	16,067
18	229,034	310,069	427,066	181,495	229,634	261,664	218,323	110,822	201,378	155,687	100,675	207,629	63,143	76,126	166,646	43,942	81,260	64,569	36,815	46,126	70,048	37,293	20,903	45,223	23,205	18,173	37,513	14,041	11,374	27,807
19	-	230,103	194,035	-	215,556	183,310	-	106,517	119,471	-	128,262	64,896	-	77,737	73,759	-	39,915	36,751	-	47,588	43,496	-	24,755	17,294	-	21,823	12,478	-	10,829	16,229
20	324,034	-	227,208	197,954	-	173,924	182,295	-	130,555	135,874	-	90,117	93,263	-	101,549	50,946	-	45,904	47,390	-	39,726	37,830	-	24,259	25,169	-	15,503	16,962	-	14,898
21	201,302	184,579	324,271	206,072	219,127	228,092	126,248	108,999	256,598	109,223	78,998	136,157	60,361	93,606	125,503	39,290	38,954	70,273	43,178	39,266	53,073	21,995	22,007	46,855	21,205	13,985	22,654	11,639	15,072	26,005
22	-	144,833	188,850	-	157,312	115,189	-	112,555	63,900	-	62,337	54,063	-	49,984	54,101	-	36,342	37,171	-	37,688	21,983	-	21,139	12,118	-	11,075	11,755	-	8,034	10,552
23	128,491	-	116,411	100,002	-	90,657	136,771	-	61,071	84,892	-	77,042	36,588	-	46,270	28,696	-	20,654	24,859	-	22,316	21,167	-	10,659	14,096	-	11,980	8,865	-	8,374
24	157,986	115,403	192,276	101,104	128,366	199,938	73,654	60,532	112,862	65,156	88,611	66,405	36,906	38,499	83,494	32,582	23,145	45,649	17,384	24,394	46,619	12,287	10,540	18,118	14,759	12,339	13,512	8,290	5,735	13,843
25	-	140,920	153,120	-	134,281	86,826	-	82,088	68,659	-	57,877	50,327	-	53,150	42,929	-	36,646	23,202	-	25,055	21,240	-	14,393	14,710	-	11,017	9,353	-	7,536	8,770
26	92,353	-	90,365	107,885	-	60,112	75,689	-	81,335	51,984	-	29,931	31,305	-	34,097	23,517	-	18,386	22,023	-	13,807	10,963	-	12,311	11,047	-	5,878	6,764	-	7,610
27	102,290	86,672	199,482	56,566	87,914	107,965	60,096	74,248	66,653	48,998	33,653	60,185	36,848	26,951	67,208	15,350	17,703	37,970	12,337	15,488	22,505	11,567	11,700	13,498	9,440	5,849	9,888	8,272	5,161	10,423
28	-	127,177	71,119	-	103,107	54,646	-	39,190	42,522	-	40,226	41,174	-	29,000	29,464	-	25,253	13,691	-	16,992	11,641	-	8,844	8,090	-	6,500	8,582	-	4,866	5,550
29	102,808	-	62,831	58,147	-	81,069	45,576	-	31,117	44,300	-	22,070	27,920	-	25,813	19,829	-	11,190	10,785	-	15,115	7,763	-	5,929	7,454	-	5,256	4,900	-	4,464
30	83,931	84,522	160,537	66,968	97,848	98,261	60,874	43,598	83,928	43,112	35,729	67,515	26,321	31,222	49,018	17,170	16,268	27,796	16,388	14,909	25,425	10,359	8,524	13,845	8,736	7,654	10,304	5,572	4,546	9,521
31	-	51,848	43,991	-	50,227	32,307	-	28,103	34,154	-	26,088	15,921	-	24,449	17,440	-	9,625	10,116	-	10,490	7,979	-	4,873	6,524	-	3,667	3,170	-	4,046	3,274
32	53,028	-	74,235	32,930	-	29,620	36,742	-	20,953	33,982	-	16,593	21,137	-	16,103	9,203	-	11,797	7,450	-	6,660	5,621	-	4,926	5,487	-	3,262	4,467	-	3,187
33	49,053	79,944	70,805	38,043	47,915	62,650	46,831	21,556	50,241	26,247	18,418	41,955	14,506	15,099	30,886	11,132	13,633	15,121	7,753	9,476	13,213	7,502	5,223	7,970	5,209	3,640	8,053	2,908	2,428	7,133
34	-	33,610	35,244	-	45,292	34,317	-	23,537	17,768	-	22,993	12,551	-	12,495	15,971	-	7,090	6,926	-	7,730	8,458	-	3,906	3,139	-	4,010	2,117	-	2,295	2,357
35	60,377	-	53,726	51,328	-	38,326	39,495	-	29,824	30,467	-	19,408	18,773	-	18,780	12,576	-	10,099	9,209	-	8,657	7,385	-	5,419	5,829	-	3,827	3,691	-	3,628
36	27,740	26,421	52,341	30,130	29,827	42,463	18,861	15,787	35,156	16,175	11,595	17,378	10,879	14,051	19,602	5,543	5,458	9,331	6,851	7,252	8,223	3,272	2,893	6,948	2,876	2,084	3,960	1,697	2,294	3,494
37	-	28,650	34,112	-	34,028	15,723	-	19,204	11,963	-	11,139	10,758	-	8,884	9,587	-	5,311	6,897	-	5,612	3,925	-	3,708	2,090	-	2,380	1,539	-	1,452	1,648
38	27,288	-	20,695	22,171	-	16,826	24,397	-	9,934	14,048	-	11,497	7,806	-	8,591	4,840	-	4,643	3,626	-	3,896	4,213	-	2,246	3,339	-	2,272	1,570	-	1,487
39	33,623	22,709	49,393	16,694	26,947	40,243	15,385	11,780	20,572	16,048	14,173	14,566	6,843	7,727	15,693	7,078	5,908	8,096	3,588	5,014	9,499	2,381	2,426	4,551	2,292	2,391	2,794	1,510	1,240	2,950
40	-	31,661	22,520	-	23,979	16,824	-	13,539	14,997	-	11,644	9,119	-	8,737	8,754	-	5,616	4,490	-	4,905	3,764	-	3,222	2,037	-	2,122	1,550	-	1,426	1,761
41	20,735	-	15,694	18,421	-	9,523	11,080	-	11,230	9,954	-	5,427	5,550	-	7,215	3,280	-	3,060	3,725	-	2,487	2,337	-	2,241	1,935	-	949	1,115	-	968
42	18,977	19,726	45,038	13,410	18,737	21,767	12,242	13,718	17,008	11,618	7,166	12,127	9,467	5,819	12,994	3,677	3,994	8,765	2,566	3,759	4,840	2,066	2,846	3,109	1,947	1,244	2,309	1,629	932	2,044
43	-	17,608	10,589	-	13,266	9,894	-	6,360	6,411	-	5,553	6,942	-	4,782	4,478	-	3,883	2,113	-	3,166	1,716	-	1,293	1,054	-	1,004	1,380	-	647	741
44	18,863	-	10,603	9,897	-	12,204	9,680	-	5,867	7,438	-	5,055	4,417	-	4,641	3,895	-	2,189	2,470	-	3,009	1,376	-	1,021	1,428	-	772	852	-	855
45	15,330	14,363	25,500	13,637	14,395	19,713	10,993	8,004	13,624	8,167	7,970	10,175	4,870	5,437	9,145	2,793	3,287	5,982	2,109	2,859	4,636	1,712	1,472	2,555	1,751	1,083	1,656	946	849	1,639
46	-	8,875	9,699	-	10,316	6,377	-	4,818	6,739	-	3,917	3,062	-	4,608	3,127	-	2,205	1,419	-	1,924	1,422	-	868	1,333	-	634	505	-	767	601
47	8,133	-	10,651	6,262	-	5,247	6,103	-	4,651	4,967	-	2,928	3,816	-	3,092	1,865	-	2,177	1,292	-	1,166	1,066	-	699	845	-	604	837	-	654
48	9,146	10,589	12,600	6,071	7,818	9,185	8,153	4,317	6,931	4,252	3,339	7,104	2,229	2,715	5,988	1,346	2,258	2,450	1,096	1,511	2,065	1,481	642	1,138	699	608	1,227	455	559	981
49	-	7,921	7,926	-	8,860	7,743	-	4,214	3,692	-	4,828	2,410	-	2,537	2,689	-	1,614	1,307	-	1,652	1,614	-	745	666	-	889	444	-	425	453
50	9,440	-	7,850	6,695	-	6,951	6,364	-	3,985	4,529	-	3,026	3,290	-	2,858	1,918	-	1,396	1,461	-	1,299	1,008	-	750	865	-	695	575	-	469

Figure 5.
First 30 x 50 tandem gaps frequencies for the first 150 million primes (the first gap represented by columns and the second by rows).

- for each a, pairs (a, b) have a uniform impossibility pattern where each third value is impossible.

A closer study reveals a more clear pattern for possible/impossible gaps:

- if either a or b is a multiple of 3 ($a = 3k$ or $b = 3k$) the pair is always possible, and,
- for the rest, if $a = 3k + i | i \in \{1, 2\}$, possible $b = 3m - i$ or $3m$; $b = 3m + i$ is always impossible.

Trying to understand the pattern, we recall that every prime number is in the form $p = 6k \pm i$—if both a and b are multiples of 3, the actual gaps sum $2a + 2b$ is a multiple of 6, so the number $p + 2(a + b)$ can be a prime.

If the first gap $a = 3m + 1$, then $p_i = 6k - 1$ always (if $p_i = 6k + 1$ then $p_{i+1} = p_i + 2a = 6k + 1 + 6m + 2 = 6r + 3 = 3(2r + 1)$, which is composite!). In this case, considering $b = 3n + i$, $p_{i+2} = p_i + 2a + 2b = 6k - 1 + 6m + 2 + 6n + 2i = 6r + 1 + 2i$—for p_{i+2} to be prime, $2i + 1$ must be either 1 or 5, so i must be 0 or 2 (if $i = 1$ then $i + 2i = 3$ and $6r + 3$ is composite!).

The same rationale is applied to $a = 3m + 2$, where $p_i = 6k + 1$ always and, in $b = 3n + i$, i must be 0 or 1, thus explaining the observed pattern.

3.1.1 *Theoretical justification of impossible tandem gaps*

Let $a, b \in \mathbb{N}$ be encoded gap values (i.e., half the actual prime gap), and consider tandem gap pairs (a, b). Empirical observations above show that certain pairs (a, b) never appear in sequences of three consecutive primes. This section provides a theoretical explanation for this phenomenon.

Let p_i be a prime, and assume the next two primes:

$$p_{i+1} = p_i + 2a, \quad p_{i+2} = p_{i+1} + 2b = p_i + 2(a + b) \tag{5}$$

It is known that all odd primes satisfy $p \equiv \pm 1 (\text{mod} 6)$, i.e., any odd prime can be written as $p = 6k \pm 1$ for some $k \in \mathbb{N}$.

Case 1: Both a and b are divisible by 3. Let $a = 3m$ and $b = 3n$. Then:

$$2a - 2b = 6(m + n) \tag{6}$$

so the total distance between p_i and p_{i+2} is divisible by 6. Since $p_i \equiv \pm 1 (\text{mod} 6)$, this shift preserves the required form for primes. Therefore, such configurations are *possible*.

Case 2: $a \equiv 1$ (mod3) Let $a = 3m + 1$. Then the prime must be of the form $p_i = 6k - 1$; otherwise, $p_{i+1} = p_i + 2a$ would not be prime. We compute:

$$p_{i+1} = 6k - 1 + 2(3m + 1) = 6k + 6m + 1 \tag{7}$$

Let $b = 3n + j$ with $j \in \{0,1,2\}$, then:

$$p_{i+2} = p_{i+1} + 2b = 6k + 6m + 1 + 2(3n + j) = 6(k + m + n) + 1 + 2j \tag{8}$$

For p_{i+2} to be prime, we must have:

$$6(k+m+n)+1+2j \equiv \pm 1 \pmod{6} \Rightarrow 2j \equiv 0 \text{ or } 4 \pmod{6} \Rightarrow j = 0 \text{ or } 2 \quad (9)$$

Hence, $b \equiv 1 \pmod{3}$ is impossible in this case.

Case 3: $a \equiv 2 \pmod{3}$ Let $a = 3m+2$, which forces $p_i = 6k+1$ (otherwise p_{i+1} is not prime). Repeating the argument above:

$$p_{i+2} = 6k+1+2(3m+2+3n+j) = 6(k+m+n)+1+2j \quad (10)$$

Again, for p_{i+2} to be prime, $2j \equiv 0$ or $4 \pmod{6}$, so $j = 0$ or 1. Therefore, $b \equiv 2 \pmod{3}$ is impossible.

Summary:

- If either $a \equiv 0 \pmod{3}$ or $b \equiv 0 \pmod{3}$, then the pair (a, b) is always *possible*.
- Otherwise:
  - If $a \equiv 1 \pmod{3}$, then $b \equiv 1 \pmod{3}$ is *impossible*.
  - If $a \equiv 2 \pmod{3}$, then $b \equiv 2 \pmod{3}$ is *impossible*.

Thus, the set of valid tandem gaps is constrained by modular congruences modulo 3 between the two components of the pair.

3.2 Symmetric pairs

Equally remarkable, if not more so, is the fact that the frequency of pairs (a, b) is essentially equal to the one of pairs (b, a), as it can be seen in **Figure 6**, where the values were normalized relative to the total number of pairs in the set (150 million) and represented in per mille with one decimal place, resulting in a practically symmetrical matrix—we have highlighted in the pictures two pairs of lines-columns to facilitate the tracking of value identities between symmetrical tandem gaps.

To further support our insights, we present in **Figures 7–10** the normalized frequencies for the first 30 x 30 tandem gaps in several sequences of 150 million consecutive primes. We can observe the aforementioned characteristics and how the percentages exhibit the same behavior as the simple gaps: frequency slowly decreasing up to 10, then increasing much more slowly above 10. The behavior is similar for sets of 100 million primes.

3.2.1. Theoretical justification

Equally remarkable is the fact that the frequency of pairs (a, b) is essentially equal to that of pairs (b, a), as illustrated in **Figure 6**. The values, normalized relative to the total number of pairs in the set (150 million) and represented in per mille, result in a practically symmetrical matrix.

While our initial empirical analysis identified this symmetry as a surprising pattern, a theoretical justification can be derived from the **Hardy-Littlewood k-tuples conjecture**.

A “tandem gap” (a, b) corresponds to a sequence of three consecutive primes of the form:

	1	2	3	4	5	6	7	8	9	10	11	12	13	14	15	16	17	18	19	20	21	22	23	24	25	26	27	28	29	30
1	-	6.7	6.2	-	8.2	6.0	-	4.1	5.2	-	3.5	3.0	-	4.2	2.8	-	1.7	1.5	-	2.2	1.3	-	0.9	1.0	-	0.6	0.7	-	0.7	0.6
2	6.7	-	9.1	5.3	-	4.5	5.5	-	4.2	4.5	-	3.4	3.7	-	3.0	1.6	-	2.1	1.5	-	1.2	1.0	-	0.8	0.9	-	0.6	0.8	-	0.6
3	6.2	9.1	10.6	6.5	7.0	8.5	7.0	3.9	6.2	4.4	3.9	6.7	2.1	2.5	4.9	1.6	2.2	2.8	1.3	1.5	2.2	1.3	0.8	1.3	1.0	0.6	1.3	0.5	0.4	1.1
4	-	5.3	6.5	-	6.4	5.2	-	3.1	3.1	-	3.9	1.9	-	2.1	2.4	-	1.5	1.2	-	1.3	1.4	-	0.7	0.7	-	0.7	0.4	-	0.4	0.4
5	8.2	-	7.0	6.4	-	6.3	6.0	-	5.1	4.4	-	3.0	2.8	-	2.8	2.2	-	1.5	1.4	-	1.5	1.1	-	0.9	0.9	-	0.6	0.7	-	0.7
6	6.0	4.5	8.4	5.2	6.3	6.1	3.8	3.5	6.8	2.9	2.4	3.3	1.8	2.7	4.4	1.1	1.1	1.7	1.2	1.2	1.5	0.8	0.6	1.3	0.6	0.4	0.7	0.4	0.5	0.7
7	-	5.5	7.0	-	6.0	3.9	-	4.3	2.4	-	2.4	2.1	-	1.9	2.2	-	1.2	1.5	-	1.2	0.8	-	0.9	0.5	-	0.5	0.4	-	0.3	0.4
8	4.1	-	3.9	3.1	-	3.5	4.3	-	2.0	2.5	-	2.1	1.6	-	1.5	0.8	-	0.7	0.7	-	0.7	0.8	-	0.4	0.6	-	0.5	0.3	-	0.3
9	5.2	4.2	6.3	3.1	5.1	6.8	2.4	2.0	3.5	2.4	2.5	3.0	1.4	1.4	2.6	1.1	0.8	1.3	0.8	0.9	1.7	0.4	0.4	0.8	0.5	0.5	0.4	0.3	0.2	0.6
10	-	4.5	4.4	-	4.4	2.9	-	2.5	2.4	-	2.5	1.8	-	1.7	1.5	-	1.0	1.0	-	0.9	0.7	-	0.6	0.4	-	0.3	0.3	-	0.3	0.3
11	3.5	-	3.9	3.9	-	2.4	2.4	-	2.5	2.5	-	1.1	1.2	-	1.3	0.7	-	0.7	0.9	-	0.5	0.4	-	0.6	0.4	-	0.2	0.3	-	0.2
12	3.0	3.4	6.7	1.9	3.0	3.3	2.1	2.1	3.0	1.8	1.1	1.8	1.3	1.0	1.9	0.7	0.6	1.4	0.4	0.6	0.9	0.4	0.5	0.4	0.3	0.2	0.4	0.3	0.1	0.4
13	-	3.7	2.1	-	2.8	1.8	-	1.6	1.4	-	1.2	1.3	-	1.0	1.2	-	0.8	0.4	-	0.6	0.4	-	0.2	0.2	-	0.2	0.2	-	0.2	0.2
14	4.2	-	2.5	2.1	-	2.7	1.9	-	1.3	1.7	-	1.0	1.0	-	1.1	0.9	-	0.5	0.5	-	0.6	0.3	-	0.3	0.4	-	0.2	0.2	-	0.2
15	2.8	3.0	4.9	2.4	2.8	4.4	2.2	1.5	2.6	1.5	1.3	1.9	1.2	1.1	1.8	0.6	0.6	1.1	0.5	0.7	0.8	0.4	0.3	0.6	0.3	0.2	0.4	0.2	0.2	0.3
16	-	1.6	1.6	-	2.2	1.1	-	0.8	1.1	-	0.7	0.7	-	0.9	0.6	-	0.4	0.3	-	0.3	0.3	-	0.2	0.2	-	0.2	0.1	-	0.1	0.1
17	1.7	-	2.2	1.5	-	1.1	1.2	-	0.9	1.0	-	0.6	0.8	-	0.6	0.4	-	0.5	0.3	-	0.3	0.2	-	0.2	0.2	-	0.1	0.2	-	0.1
18	1.5	2.1	2.8	1.2	1.5	1.7	1.5	0.7	1.3	1.0	0.7	1.4	0.4	0.5	1.1	0.3	0.5	0.4	0.2	0.3	0.5	0.2	0.1	0.3	0.2	0.1	0.3	0.1	0.1	0.2
19	-	1.5	1.3	-	1.4	1.2	-	0.7	0.8	-	0.9	0.4	-	0.5	0.5	-	0.3	0.2	-	0.3	0.3	-	0.2	0.1	-	0.1	0.1	-	0.1	0.1
20	2.2	-	1.5	1.3	-	1.2	1.2	-	0.9	0.9	-	0.6	0.6	-	0.7	0.3	-	0.3	0.3	-	0.3	0.3	-	0.2	0.2	-	0.1	0.1	-	0.1
21	1.3	1.2	2.2	1.4	1.5	1.5	0.8	0.7	1.7	0.7	0.5	0.9	0.4	0.6	0.8	0.3	0.3	0.5	0.3	0.3	0.4	0.1	0.1	0.3	0.1	0.1	0.2	0.1	0.1	0.2
22	-	1.0	1.3	-	1.0	0.8	-	0.8	0.4	-	0.4	0.4	-	0.3	0.4	-	0.2	0.2	-	0.3	0.1	-	0.1	0.1	-	0.1	0.1	-	0.1	0.1
23	0.9	-	0.8	0.7	-	0.6	0.9	-	0.4	0.6	-	0.5	0.2	-	0.3	0.2	-	0.1	0.2	-	0.1	0.1	-	0.1	0.1	-	0.1	0.1	-	0.1
24	1.1	0.8	1.3	0.7	0.9	1.3	0.5	0.4	0.8	0.4	0.6	0.4	0.2	0.3	0.6	0.2	0.2	0.3	0.1	0.2	0.3	0.1	0.1	0.1	0.1	0.1	0.1	0.1	0.0	0.1
25	-	0.9	1.0	-	0.9	0.6	-	0.5	0.5	-	0.4	0.3	-	0.4	0.3	-	0.2	0.2	-	0.2	0.1	-	0.1	0.1	-	0.1	0.1	-	0.1	0.1
26	0.6	-	0.6	0.7	-	0.4	0.5	-	0.5	0.3	-	0.2	0.2	-	0.2	0.2	-	0.1	0.1	-	0.1	0.1	-	0.1	0.1	-	0.0	0.0	-	0.1
27	0.7	0.6	1.3	0.4	0.6	0.7	0.4	0.5	0.4	0.3	0.2	0.4	0.2	0.2	0.4	0.1	0.1	0.3	0.1	0.1	0.2	0.1	0.1	0.1	0.1	0.0	0.1	0.1	0.0	0.1
28	-	0.8	0.5	-	0.7	0.4	-	0.3	0.3	-	0.3	0.3	-	0.2	0.2	-	0.2	0.1	-	0.1	0.1	-	0.1	0.1	-	0.0	0.1	-	0.0	0.0
29	0.7	-	0.4	0.4	-	0.5	0.3	-	0.2	0.3	-	0.1	0.2	-	0.2	0.1	-	0.1	0.1	-	0.1	0.1	-	0.0	0.0	-	0.0	0.0	-	0.0
30	0.6	0.6	1.1	0.4	0.7	0.7	0.4	0.3	0.6	0.3	0.2	0.5	0.2	0.2	0.3	0.1	0.1	0.2	0.1	0.1	0.2	0.1	0.1	0.1	0.1	0.1	0.1	0.0	0.0	0.1
31	-	0.3	0.3	-	0.3	0.2	-	0.2	0.2	-	0.2	0.1	-	0.2	0.1	-	0.1	0.1	-	0.1	0.1	-	0.0	0.0	-	0.0	0.0	-	0.0	0.0
32	0.4	-	0.5	0.2	-	0.2	0.2	-	0.1	0.2	-	0.1	0.1	-	0.1	0.1	-	0.1	0.0	-	0.0	0.0	-	0.0	0.0	-	0.0	0.0	-	0.0
33	0.3	0.5	0.5	0.3	0.3	0.4	0.3	0.1	0.3	0.2	0.1	0.3	0.1	0.1	0.2	0.1	0.1	0.1	0.1	0.1	0.1	0.1	0.0	0.1	0.0	0.0	0.1	0.0	0.0	0.0
34	-	0.2	0.2	-	0.3	0.2	-	0.2	0.1	-	0.2	0.1	-	0.1	0.1	-	0.0	0.0	-	0.1	0.1	-	0.0	0.0	-	0.0	0.0	-	0.0	0.0
35	0.4	-	0.4	0.3	-	0.3	0.3	-	0.2	0.2	-	0.1	0.1	-	0.1	0.1	-	0.1	0.1	-	0.1	0.0	-	0.0	0.0	-	0.0	0.0	-	0.0
36	0.2	0.2	0.3	0.2	0.2	0.3	0.1	0.1	0.2	0.1	0.1	0.1	0.1	0.1	0.1	0.0	0.0	0.1	0.0	0.0	0.1	0.0	0.0	0.0	0.0	0.0	0.0	0.0	0.0	0.0
37	-	0.2	0.2	-	0.2	0.1	-	0.1	0.1	-	0.1	0.1	-	0.1	0.1	-	0.0	0.0	-	0.0	0.0	-	0.0	0.0	-	0.0	0.0	-	0.0	0.0
38	0.2	-	0.1	0.1	-	0.1	0.2	-	0.1	0.1	-	0.1	0.1	-	0.1	0.0	-	0.0	0.0	-	0.0	0.0	-	0.0	0.0	-	0.0	0.0	-	0.0
39	0.2	0.2	0.3	0.1	0.2	0.3	0.1	0.1	0.1	0.1	0.1	0.1	0.0	0.1	0.1	0.0	0.0	0.1	0.0	0.0	0.1	0.0	0.0	0.0	0.0	0.0	0.0	0.0	0.0	0.0
40	-	0.2	0.2	-	0.2	0.1	-	0.1	0.1	-	0.1	0.1	-	0.1	0.1	-	0.0	0.0	-	0.0	0.0	-	0.0	0.0	-	0.0	0.0	-	0.0	0.0
41	0.1	-	0.1	0.1	-	0.1	0.1	-	0.1	0.1	-	0.0	0.0	-	0.0	0.0	-	0.0	0.0	-	0.0	0.0	-	0.0	0.0	-	0.0	0.0	-	0.0
42	0.1	0.1	0.3	0.1	0.1	0.1	0.1	0.1	0.1	0.1	0.0	0.1	0.1	0.0	0.1	0.0	0.0	0.1	0.0	0.0	0.0	0.0	0.0	0.0	0.0	0.0	0.0	0.0	0.0	0.0
43	-	0.1	0.1	-	0.1	0.1	-	0.0	0.0	-	0.0	0.0	-	0.0	0.0	-	0.0	0.0	-	0.0	0.0	-	0.0	0.0	-	0.0	0.0	-	0.0	0.0
44	0.1	-	0.1	0.1	-	0.1	0.1	-	0.0	0.0	-	0.0	0.0	-	0.0	0.0	-	0.0	0.0	-	0.0	0.0	-	0.0	0.0	-	0.0	0.0	-	0.0
45	0.1	0.1	0.2	0.1	0.1	0.1	0.1	0.1	0.1	0.1	0.1	0.1	0.0	0.0	0.1	0.0	0.0	0.0	0.0	0.0	0.0	0.0	0.0	0.0	0.0	0.0	0.0	0.0	0.0	0.0
46	-	0.1	0.1	-	0.1	0.0	-	0.0	0.0	-	0.0	0.0	-	0.0	0.0	-	0.0	0.0	-	0.0	0.0	-	0.0	0.0	-	0.0	0.0	-	0.0	0.0
47	0.1	-	0.1	0.0	-	0.0	0.0	-	0.0	0.0	-	0.0	0.0	-	0.0	0.0	-	0.0	0.0	-	0.0	0.0	-	0.0	0.0	-	0.0	0.0	-	0.0
48	0.1	0.1	0.1	0.0	0.1	0.1	0.1	0.0	0.0	0.0	0.0	0.0	0.0	0.0	0.0	0.0	0.0	0.0	0.0	0.0	0.0	0.0	0.0	0.0	0.0	0.0	0.0	0.0	0.0	0.0
49	-	0.1	0.1	-	0.1	0.1	-	0.0	0.0	-	0.0	0.0	-	0.0	0.0	-	0.0	0.0	-	0.0	0.0	-	0.0	0.0	-	0.0	0.0	-	0.0	0.0
50	0.1	-	0.1	0.0	-	0.0	0.0	-	0.0	0.0	-	0.0	0.0	-	0.0	0.0	-	0.0	0.0	-	0.0	0.0	-	0.0	0.0	-	0.0	0.0	-	0.0

Figure 6.
First 30 x 50 tandem gaps, normalized [%], for the first 150 million primes.

	1	2	3	4	5	6	7	8	9	10	11	12	13	14	15	16	17	18	19	20	21	22	23	24	25	26	27	28	29	30
1	-	6.6	6.2	-	8.1	6.0	-	4.0	5.2	-	3.5	3.0	-	4.2	2.8	-	1.7	1.5	-	2.2	1.3	-	0.9	1.1	-	0.6	0.7	-	0.7	0.6
2	6.6	-	9.0	5.3	-	4.5	5.5	-	4.2	4.5	-	3.4	3.7	-	3.0	1.6	-	2.1	1.5	-	1.2	1.0	-	0.8	0.9	-	0.6	0.8	-	0.6
3	6.2	9.0	10.6	6.5	7.0	8.4	6.9	3.9	6.2	4.4	3.9	6.6	2.1	2.5	4.9	1.6	2.2	2.8	1.3	1.5	2.2	1.3	0.8	1.3	1.0	0.6	1.3	0.5	0.4	1.1
4	-	5.3	6.5	-	6.4	5.2	-	3.1	3.1	-	3.9	1.9	-	2.1	2.4	-	1.6	1.2	-	1.3	1.4	-	0.7	0.7	-	0.7	0.4	-	0.4	0.4
5	8.1	-	6.9	6.4	-	6.3	6.0	-	5.0	4.4	-	3.0	2.8	-	2.8	2.2	-	1.5	1.4	-	1.5	1.1	-	0.9	0.9	-	0.6	0.7	-	0.7
6	6.0	4.5	8.4	5.2	6.3	6.1	3.8	3.5	6.8	2.9	2.4	3.3	1.8	2.7	4.4	1.1	1.1	1.7	1.2	1.2	1.5	0.8	0.6	1.3	0.6	0.4	0.7	0.4	0.5	0.7
7	-	5.5	6.9	-	5.9	3.8	-	4.3	2.4	-	2.4	2.1	-	1.9	2.2	-	1.2	1.5	-	1.2	0.8	-	0.9	0.5	-	0.5	0.4	-	0.3	0.4
8	4.0	-	3.9	3.1	-	3.5	4.3	-	2.0	2.5	-	2.1	1.6	-	1.5	0.8	-	0.7	0.7	-	0.7	0.8	-	0.4	0.6	-	0.5	0.3	-	0.3
9	5.2	4.2	6.2	3.1	5.0	6.8	2.4	2.0	3.5	2.4	2.5	3.0	1.4	1.4	2.6	1.2	0.9	1.3	0.8	0.9	1.7	0.4	0.4	0.8	0.5	0.5	0.4	0.3	0.2	0.6
10	-	4.5	4.3	-	4.4	2.9	-	2.5	2.4	-	2.5	1.8	-	1.7	1.5	-	1.0	1.0	-	0.9	0.7	-	0.6	0.4	-	0.3	0.3	-	0.3	0.3
11	3.5	-	3.9	3.9	-	2.4	2.4	-	2.5	2.5	-	1.1	1.2	-	1.3	0.7	-	0.7	0.9	-	0.5	0.4	-	0.6	0.4	-	0.2	0.3	-	0.2
12	3.0	3.4	6.7	1.9	3.0	3.3	2.1	2.1	3.0	1.8	1.1	1.8	1.3	1.0	1.9	0.7	0.7	1.4	0.4	0.6	0.9	0.4	0.5	0.4	0.3	0.2	0.4	0.3	0.1	0.5
13	-	3.7	2.1	-	2.8	1.8	-	1.6	1.4	-	1.2	1.3	-	1.0	1.2	-	0.8	0.4	-	0.6	0.4	-	0.2	0.2	-	0.2	0.2	-	0.2	0.2
14	4.2	-	2.5	2.1	-	2.7	1.9	-	1.3	1.7	-	1.0	1.0	-	1.1	0.9	-	0.5	0.5	-	0.6	0.3	-	0.3	0.4	-	0.2	0.2	-	0.2
15	2.8	3.0	4.9	2.4	2.8	4.4	2.2	1.5	2.6	1.5	1.3	1.9	1.2	1.1	1.8	0.6	0.6	1.1	0.5	0.7	0.8	0.4	0.3	0.6	0.3	0.2	0.4	0.2	0.2	0.3
16	-	1.6	1.6	-	2.2	1.1	-	0.8	1.2	-	0.7	0.7	-	0.9	0.6	-	0.4	0.3	-	0.3	0.3	-	0.2	0.2	-	0.2	0.1	-	0.1	0.1
17	1.7	-	2.2	1.5	-	1.1	1.2	-	0.9	1.0	-	0.6	0.8	-	0.6	0.4	-	0.5	0.3	-	0.3	0.2	-	0.2	0.2	-	0.1	0.2	-	0.1
18	1.5	2.1	2.9	1.2	1.5	1.7	1.5	0.7	1.3	1.0	0.7	1.4	0.4	0.5	1.1	0.3	0.5	0.4	0.2	0.3	0.5	0.2	0.1	0.3	0.2	0.1	0.3	0.1	0.1	0.2
19	-	1.5	1.3	-	1.4	1.2	-	0.7	0.8	-	0.9	0.4	-	0.5	0.5	-	0.3	0.2	-	0.3	0.3	-	0.2	0.1	-	0.1	0.1	-	0.1	0.1
20	2.2	-	1.5	1.3	-	1.2	1.2	-	0.9	0.9	-	0.6	0.6	-	0.7	0.3	-	0.3	0.3	-	0.3	0.3	-	0.2	0.2	-	0.1	0.1	-	0.1
21	1.3	1.2	2.2	1.4	1.5	1.5	0.8	0.7	1.7	0.7	0.5	0.9	0.4	0.6	0.8	0.3	0.3	0.5	0.3	0.3	0.4	0.1	0.1	0.3	0.1	0.1	0.2	0.1	0.1	0.2
22	-	1.0	1.3	-	1.1	0.8	-	0.8	0.4	-	0.4	0.4	-	0.3	0.4	-	0.2	0.2	-	0.3	0.1	-	0.1	0.1	-	0.1	0.1	-	0.1	0.1
23	0.9	-	0.8	0.7	-	0.6	0.9	-	0.4	0.6	-	0.5	0.2	-	0.3	0.2	-	0.1	0.2	-	0.1	0.1	-	0.1	0.1	-	0.1	0.1	-	0.1
24	1.1	0.8	1.3	0.7	0.9	1.3	0.5	0.4	0.8	0.4	0.6	0.4	0.2	0.3	0.6	0.2	0.2	0.3	0.1	0.2	0.3	0.1	0.1	0.1	0.1	0.1	0.1	0.1	0.0	0.1
25	-	0.9	1.0	-	0.9	0.6	-	0.5	0.5	-	0.4	0.3	-	0.4	0.3	-	0.2	0.2	-	0.2	0.1	-	0.1	0.1	-	0.1	0.1	-	0.1	0.1
26	0.6	-	0.6	0.7	-	0.4	0.5	-	0.5	0.3	-	0.2	0.2	-	0.2	0.2	-	0.1	0.1	-	0.1	0.1	-	0.1	0.1	-	0.0	0.0	-	0.1
27	0.7	0.6	1.3	0.4	0.6	0.7	0.4	0.5	0.4	0.3	0.2	0.4	0.2	0.2	0.5	0.1	0.1	0.3	0.1	0.1	0.2	0.1	0.1	0.1	0.1	0.0	0.1	0.1	0.0	0.1
28	-	0.9	0.5	-	0.7	0.4	-	0.3	0.3	-	0.3	0.3	-	0.2	0.2	-	0.2	0.1	-	0.1	0.1	-	0.1	0.1	-	0.0	0.1	-	0.0	0.0
29	0.7	-	0.4	0.4	-	0.5	0.3	-	0.2	0.3	-	0.1	0.2	-	0.2	0.1	-	0.1	0.1	-	0.1	0.1	-	0.0	0.1	-	0.0	0.0	-	0.0
30	0.6	0.6	1.1	0.4	0.7	0.7	0.4	0.3	0.6	0.3	0.2	0.5	0.2	0.2	0.3	0.1	0.1	0.2	0.1	0.1	0.2	0.1	0.1	0.1	0.1	0.1	0.1	0.0	0.0	0.1

Figure 7.
First 30x30 tandem gaps, normalized [%], for the first 150 million primes after 1 million.

	1	2	3	4	5	6	7	8	9	10	11	12	13	14	15	16	17	18	19	20	21	22	23	24	25	26	27	28	29	30
1	-	6.6	6.1	-	8.1	5.9	-	4.0	5.2	-	3.5	3.0	-	4.2	2.8	-	1.7	1.5	-	2.2	1.4	-	0.9	1.1	-	0.6	0.7	-	0.7	0.6
2	6.6	-	9.0	5.3	-	4.5	5.5	-	4.2	4.5	-	3.4	3.7	-	3.0	1.6	-	2.1	1.5	-	1.2	1.0	-	0.8	0.9	-	0.6	0.8	-	0.6
3	6.1	9.0	10.5	6.5	6.9	8.4	6.9	3.9	6.2	4.3	3.9	6.6	2.1	2.5	4.9	1.6	2.2	2.8	1.3	1.5	2.2	1.3	0.8	1.3	1.0	0.6	1.3	0.5	0.4	1.1
4	-	5.3	6.5	-	6.3	5.2	-	3.1	3.1	-	3.9	1.9	-	2.1	2.4	-	1.6	1.2	-	1.3	1.4	-	0.7	0.7	-	0.7	0.4	-	0.4	0.4
5	8.1	-	6.9	6.3	-	6.3	5.9	-	5.0	4.4	-	3.0	2.8	-	2.8	2.2	-	1.5	1.4	-	1.5	1.1	-	0.9	0.9	-	0.6	0.7	-	0.7
6	5.9	4.5	8.4	5.2	6.3	6.1	3.8	3.5	6.8	2.9	2.4	3.3	1.8	2.7	4.4	1.1	1.1	1.8	1.2	1.2	1.5	0.8	0.6	1.3	0.6	0.4	0.7	0.4	0.5	0.7
7	-	5.5	6.9	-	5.9	3.8	-	4.3	2.4	-	2.4	2.1	-	1.9	2.2	-	1.2	1.5	-	1.2	0.8	-	0.9	0.5	-	0.5	0.4	-	0.3	0.4
8	4.0	-	3.9	3.1	-	3.4	4.3	-	2.0	2.5	-	2.1	1.6	-	1.5	0.8	-	0.7	0.7	-	0.7	0.8	-	0.4	0.6	-	0.5	0.3	-	0.3
9	5.2	4.2	6.2	3.1	5.0	6.8	2.4	2.0	3.5	2.4	2.5	3.0	1.4	1.4	2.6	1.2	0.9	1.3	0.8	0.9	1.7	0.4	0.4	0.8	0.5	0.5	0.4	0.3	0.2	0.6
10	-	4.5	4.3	-	4.4	2.9	-	2.5	2.4	-	2.5	1.8	-	1.7	1.5	-	1.0	1.0	-	0.9	0.7	-	0.6	0.4	-	0.3	0.3	-	0.3	0.3
11	3.5	-	3.9	3.9	-	2.4	2.4	-	2.5	2.5	-	1.1	1.2	-	1.3	0.7	-	0.7	0.9	-	0.5	0.4	-	0.6	0.4	-	0.2	0.3	-	0.2
12	3.0	3.4	6.7	1.9	3.0	3.3	2.1	2.1	3.0	1.8	1.1	1.8	1.3	1.0	1.9	0.7	0.7	1.4	0.4	0.6	0.9	0.4	0.5	0.4	0.3	0.2	0.4	0.3	0.1	0.5
13	-	3.7	2.1	-	2.8	1.8	-	1.6	1.4	-	1.2	1.3	-	1.0	1.2	-	0.8	0.4	-	0.6	0.4	-	0.2	0.2	-	0.2	0.3	-	0.2	0.2
14	4.2	-	2.5	2.1	-	2.7	1.9	-	1.3	1.7	-	1.0	1.0	-	1.1	0.9	-	0.5	0.5	-	0.6	0.3	-	0.3	0.4	-	0.2	0.2	-	0.2
15	2.8	3.0	4.9	2.4	2.8	4.4	2.2	1.5	2.6	1.5	1.3	1.9	1.2	1.1	1.8	0.6	0.7	1.1	0.5	0.7	0.8	0.4	0.3	0.6	0.3	0.2	0.5	0.2	0.2	0.3
16	-	1.6	1.6	-	2.2	1.1	-	0.8	1.2	-	0.7	0.7	-	0.9	0.6	-	0.4	0.3	-	0.3	0.3	-	0.2	0.2	-	0.2	0.1	-	0.1	0.1
17	1.7	-	2.2	1.5	-	1.1	1.2	-	0.9	1.0	-	0.6	0.8	-	0.7	0.4	-	0.5	0.3	-	0.3	0.2	-	0.2	0.2	-	0.1	0.2	-	0.1
18	1.5	2.1	2.9	1.2	1.5	1.7	1.5	0.7	1.3	1.0	0.7	1.4	0.4	0.5	1.1	0.3	0.5	0.4	0.2	0.3	0.5	0.3	0.1	0.3	0.2	0.1	0.3	0.1	0.1	0.2
19	-	1.5	1.3	-	1.4	1.2	-	0.7	0.8	-	0.9	0.4	-	0.5	0.5	-	0.3	0.2	-	0.3	0.3	-	0.2	0.1	-	0.1	0.1	-	0.1	0.1
20	2.2	-	1.5	1.3	-	1.2	1.2	-	0.9	0.9	-	0.6	0.6	-	0.7	0.3	-	0.3	0.3	-	0.3	0.3	-	0.2	0.2	-	0.1	0.1	-	0.1
21	1.3	1.2	2.2	1.4	1.5	1.5	0.8	0.7	1.7	0.7	0.5	0.9	0.4	0.6	0.8	0.3	0.3	0.5	0.3	0.3	0.4	0.1	0.1	0.3	0.1	0.1	0.2	0.1	0.1	0.2
22	-	1.0	1.3	-	1.1	0.8	-	0.8	0.4	-	0.4	0.4	-	0.3	0.4	-	0.2	0.2	-	0.3	0.1	-	0.1	0.1	-	0.1	0.1	-	0.1	0.1
23	0.9	-	0.8	0.7	-	0.6	0.9	-	0.4	0.6	-	0.5	0.2	-	0.3	0.2	-	0.1	0.2	-	0.2	0.1	-	0.1	0.1	-	0.1	0.1	-	0.1
24	1.1	0.8	1.3	0.7	0.9	1.3	0.5	0.4	0.8	0.4	0.6	0.4	0.2	0.3	0.6	0.2	0.2	0.3	0.1	0.2	0.3	0.1	0.1	0.1	0.1	0.1	0.1	0.1	0.0	0.1
25	-	0.9	1.0	-	0.9	0.6	-	0.6	0.5	-	0.4	0.3	-	0.4	0.3	-	0.2	0.2	-	0.2	0.1	-	0.1	0.1	-	0.1	0.1	-	0.1	0.1
26	0.6	-	0.6	0.7	-	0.4	0.5	-	0.5	0.3	-	0.2	0.2	-	0.2	0.2	-	0.1	0.1	-	0.1	0.1	-	0.1	0.1	-	0.0	0.0	-	0.1
27	0.7	0.6	1.3	0.4	0.6	0.7	0.4	0.5	0.4	0.3	0.2	0.4	0.2	0.2	0.5	0.1	0.1	0.3	0.1	0.1	0.2	0.1	0.1	0.1	0.1	0.0	0.1	0.1	0.0	0.1
28	-	0.9	0.5	-	0.7	0.4	-	0.3	0.3	-	0.3	0.3	-	0.2	0.2	-	0.2	0.1	-	0.1	0.1	-	0.1	0.1	-	0.0	0.1	-	0.0	0.0
29	0.7	-	0.4	0.4	-	0.5	0.3	-	0.2	0.3	-	0.1	0.2	-	0.2	0.1	-	0.1	0.1	-	0.1	0.1	-	0.0	0.1	-	0.0	0.0	-	0.0
30	0.6	0.6	1.1	0.4	0.7	0.7	0.4	0.3	0.6	0.3	0.2	0.5	0.2	0.2	0.3	0.1	0.1	0.2	0.1	0.1	0.2	0.1	0.1	0.1	0.1	0.1	0.1	0.0	0.0	0.1

Figure 8.
First 30x30 tandem gaps, normalized [%], for the first 150 million primes after 2 million.

	1	2	3	4	5	6	7	8	9	10	11	12	13	14	15	16	17	18	19	20	21	22	23	24	25	26	27	28	29	30
1	-	6.1	5.7	-	7.6	5.6	-	3.8	5.0	-	3.3	2.9	-	4.1	2.7	-	1.7	1.5	-	2.2	1.4	-	0.9	1.1	-	0.6	0.7	-	0.7	0.6
2	6.1	-	8.4	4.9	-	4.2	5.2	-	4.0	4.4	-	3.2	3.6	-	2.9	1.6	-	2.1	1.5	-	1.3	1.0	-	0.8	1.0	-	0.6	0.9	-	0.6
3	5.7	8.4	9.9	6.1	6.5	8.0	6.6	3.7	6.0	4.2	3.8	6.5	2.1	2.5	4.8	1.5	2.2	2.8	1.3	1.5	2.2	1.3	0.8	1.3	1.1	0.6	1.4	0.5	0.4	1.1
4	-	4.9	6.1	-	6.0	5.0	-	3.0	3.0	-	3.8	1.9	-	2.1	2.3	-	1.6	1.2	-	1.3	1.4	-	0.7	0.7	-	0.8	0.4	-	0.4	0.5
5	7.6	-	6.5	6.0	-	6.0	5.7	-	4.9	4.3	-	2.9	2.8	-	2.8	2.2	-	1.6	1.5	-	1.5	1.1	-	0.9	0.9	-	0.6	0.7	-	0.7
6	5.6	4.3	8.0	5.0	6.0	5.8	3.7	3.3	6.6	2.8	2.3	3.3	1.8	2.7	4.4	1.1	1.1	1.8	1.2	1.2	1.6	0.8	0.6	1.4	0.6	0.4	0.8	0.4	0.6	0.7
7	-	5.2	6.6	-	5.7	3.7	-	4.1	2.4	-	2.4	2.1	-	1.9	2.2	-	1.3	1.5	-	1.2	0.9	-	1.0	0.5	-	0.5	0.4	-	0.3	0.4
8	3.8	-	3.7	3.0	-	3.3	4.1	-	2.0	2.4	-	2.1	1.6	-	1.5	0.9	-	0.8	0.7	-	0.8	0.8	-	0.4	0.6	-	0.5	0.3	-	0.3
9	5.0	4.0	6.0	3.0	4.9	6.6	2.4	2.0	3.4	2.4	2.5	3.0	1.4	1.4	2.7	1.2	0.9	1.4	0.8	0.9	1.8	0.4	0.4	0.8	0.5	0.6	0.5	0.3	0.2	0.6
10	-	4.3	4.2	-	4.3	2.8	-	2.4	2.4	-	2.5	1.8	-	1.7	1.5	-	1.1	1.1	-	0.9	0.8	-	0.6	0.5	-	0.4	0.4	-	0.3	0.3
11	3.3	-	3.8	3.8	-	2.3	2.4	-	2.5	2.5	-	1.1	1.2	-	1.4	0.7	-	0.7	0.9	-	0.6	0.4	-	0.6	0.4	-	0.2	0.3	-	0.3
12	2.9	3.2	6.5	1.9	2.9	3.3	2.1	2.1	3.0	1.8	1.1	1.8	1.3	1.0	2.0	0.7	0.7	1.4	0.5	0.6	1.0	0.4	0.6	0.5	0.4	0.2	0.4	0.3	0.2	0.5
13	-	3.6	2.1	-	2.8	1.8	-	1.6	1.4	-	1.2	1.3	-	1.0	1.2	-	0.9	0.4	-	0.7	0.4	-	0.3	0.3	-	0.2	0.3	-	0.2	0.2
14	4.1	-	2.5	2.1	-	2.7	1.9	-	1.4	1.7	-	1.0	1.0	-	1.1	0.9	-	0.5	0.5	-	0.7	0.4	-	0.3	0.4	-	0.2	0.2	-	0.2
15	2.7	2.9	4.8	2.3	2.8	4.4	2.2	1.5	2.7	1.5	1.4	2.0	1.2	1.1	1.9	0.6	0.7	1.2	0.5	0.7	0.9	0.4	0.3	0.6	0.3	0.3	0.5	0.2	0.2	0.4
16	-	1.6	1.5	-	2.2	1.1	-	0.9	1.2	-	0.7	0.7	-	0.9	0.6	-	0.4	0.3	-	0.4	0.3	-	0.2	0.2	-	0.2	0.1	-	0.1	0.1
17	1.7	-	2.2	1.6	-	1.1	1.3	-	0.9	1.1	-	0.7	0.9	-	0.7	0.4	-	0.6	0.3	-	0.3	0.3	-	0.2	0.3	-	0.1	0.2	-	0.1
18	1.5	2.1	2.9	1.2	1.5	1.8	1.5	0.8	1.4	1.1	0.7	1.4	0.4	0.5	1.2	0.3	0.6	0.5	0.3	0.3	0.5	0.3	0.2	0.3	0.2	0.1	0.3	0.1	0.1	0.2
19	-	1.5	1.3	-	1.5	1.2	-	0.7	0.8	-	0.9	0.5	-	0.5	0.5	-	0.3	0.3	-	0.3	0.3	-	0.2	0.1	-	0.2	0.1	-	0.1	0.1
20	2.2	-	1.5	1.3	-	1.2	1.2	-	0.9	0.9	-	0.6	0.7	-	0.7	0.4	-	0.3	0.3	-	0.3	0.3	-	0.2	0.2	-	0.1	0.1	-	0.1
21	1.4	1.2	2.2	1.4	1.5	1.6	0.9	0.8	1.8	0.8	0.6	1.0	0.4	0.7	0.9	0.3	0.3	0.5	0.3	0.3	0.4	0.2	0.2	0.3	0.2	0.1	0.2	0.1	0.1	0.2
22	-	1.0	1.3	-	1.1	0.8	-	0.8	0.4	-	0.4	0.4	-	0.4	0.4	-	0.3	0.3	-	0.3	0.2	-	0.2	0.1	-	0.1	0.1	-	0.1	0.1
23	0.9	-	0.8	0.7	-	0.6	1.0	-	0.4	0.6	-	0.5	0.3	-	0.3	0.2	-	0.2	0.2	-	0.2	0.2	-	0.1	0.1	-	0.1	0.1	-	0.1
24	1.1	0.8	1.3	0.7	0.9	1.4	0.5	0.4	0.8	0.5	0.6	0.5	0.3	0.3	0.6	0.2	0.2	0.3	0.1	0.2	0.3	0.1	0.1	0.1	0.1	0.1	0.1	0.1	0.0	0.1
25	-	1.0	1.1	-	0.9	0.6	-	0.6	0.5	-	0.4	0.4	-	0.4	0.3	-	0.3	0.2	-	0.2	0.2	-	0.1	0.1	-	0.1	0.1	-	0.1	0.1
26	0.6	-	0.6	0.8	-	0.4	0.5	-	0.6	0.4	-	0.2	0.2	-	0.3	0.2	-	0.1	0.2	-	0.1	0.1	-	0.1	0.1	-	0.0	0.1	-	0.1
27	0.7	0.6	1.4	0.4	0.6	0.8	0.4	0.5	0.5	0.4	0.2	0.4	0.3	0.2	0.5	0.1	0.1	0.3	0.1	0.1	0.2	0.1	0.1	0.1	0.1	0.0	0.1	0.1	0.0	0.1
28	-	0.9	0.5	-	0.7	0.4	-	0.3	0.3	-	0.3	0.3	-	0.2	0.2	-	0.2	0.1	-	0.1	0.1	-	0.1	0.1	-	0.1	0.1	-	0.0	0.0
29	0.7	-	0.4	0.4	-	0.6	0.3	-	0.2	0.3	-	0.2	0.2	-	0.2	0.1	-	0.1	0.1	-	0.1	0.1	-	0.0	0.1	-	0.0	0.0	-	0.0
30	0.6	0.6	1.1	0.5	0.7	0.7	0.4	0.3	0.6	0.3	0.3	0.5	0.2	0.2	0.4	0.1	0.1	0.2	0.1	0.1	0.2	0.1	0.1	0.1	0.1	0.1	0.1	0.0	0.0	0.1

Figure 9.
First 30x30 tandem gaps, normalized [%], for the first 150 million primes after 50 million.

$$p, \quad p + 2a, \quad p + 2a + 2b \tag{11}$$

This configuration is associated with the admissible tuple (or constellation) $\mathcal{H}_1 = \{0, 2a, 2(a+b)\}$.

Conversely, the symmetric tandem gap (b, a) corresponds to the sequence:

$$q, \quad q + 2b, \quad q + 2b + 2a \tag{12}$$

associated with the tuple $\mathcal{H}_2 = \{0, 2b, 2(b+a)\}$.

According to the Hardy-Littlewood conjecture, the asymptotic density of prime numbers matching a constellation $\mathcal{H}$ is proportional to a singular series $\mathfrak{S}(\mathcal{H})$, which

	1	2	3	4	5	6	7	8	9	10	11	12	13	14	15	16	17	18	19	20	21	22	23	24	25	26	27	28	29	30
1	-	6.1	5.7	-	7.6	5.6	-	3.8	5.0	-	3.3	2.9	-	4.1	2.7	-	1.7	1.5	-	2.2	1.4	-	0.9	1.1	-	0.6	0.7	-	0.7	0.6
2	6.1	-	8.4	4.9	-	4.2	5.2	-	4.0	4.4	-	3.2	3.6	-	2.9	1.6	-	2.1	1.5	-	1.3	1.0	-	0.8	1.0	-	0.6	0.9	-	0.6
3	5.7	8.4	9.9	6.1	6.5	8.0	6.6	3.7	6.0	4.2	3.8	6.5	2.1	2.5	4.8	1.5	2.2	2.8	1.3	1.5	2.2	1.3	0.8	1.3	1.1	0.6	1.4	0.5	0.4	1.1
4	-	4.9	6.1	-	6.0	5.0	-	3.0	3.0	-	3.8	1.9	-	2.1	2.3	-	1.6	1.2	-	1.3	1.4	-	0.7	0.7	-	0.8	0.4	-	0.4	0.5
5	7.6	-	6.5	6.0	-	5.9	5.7	-	4.9	4.3	-	2.9	2.8	-	2.8	2.2	-	1.6	1.5	-	1.5	1.1	-	0.9	0.9	-	0.6	0.7	-	0.7
6	5.6	4.2	7.9	5.0	6.0	5.8	3.7	3.3	6.6	2.8	2.3	3.3	1.8	2.7	4.4	1.1	1.1	1.8	1.2	1.2	1.6	0.8	0.6	1.4	0.6	0.4	0.8	0.4	0.6	0.7
7	-	5.2	6.6	-	5.7	3.7	-	4.1	2.4	-	2.4	2.1	-	1.9	2.2	-	1.3	1.5	-	1.2	0.9	-	1.0	0.5	-	0.5	0.4	-	0.3	0.4
8	3.8	-	3.7	3.0	-	3.3	4.1	-	2.0	2.4	-	2.1	1.6	-	1.5	0.9	-	0.8	0.7	-	0.8	0.8	-	0.4	0.6	-	0.5	0.3	-	0.3
9	5.0	4.0	6.0	3.0	4.9	6.6	2.4	2.0	3.4	2.4	2.5	3.0	1.4	1.4	2.7	1.2	0.9	1.4	0.8	0.9	1.8	0.4	0.4	0.8	0.5	0.6	0.5	0.3	0.2	0.6
10	-	4.3	4.2	-	4.3	2.8	-	2.4	2.4	-	2.5	1.8	-	1.7	1.5	-	1.1	1.1	-	0.9	0.8	-	0.6	0.5	-	0.4	0.4	-	0.3	0.3
11	3.3	-	3.8	3.8	-	2.3	2.4	-	2.5	2.5	-	1.1	1.2	-	1.4	0.7	-	0.7	0.9	-	0.6	0.4	-	0.6	0.4	-	0.2	0.3	-	0.3
12	2.9	3.2	6.5	1.9	2.9	3.3	2.1	2.1	3.0	1.8	1.1	1.8	1.3	1.0	2.0	0.7	0.7	1.4	0.5	0.6	1.0	0.4	0.6	0.5	0.4	0.2	0.4	0.3	0.2	0.5
13	-	3.6	2.1	-	2.8	1.8	-	1.6	1.4	-	1.2	1.3	-	1.0	1.2	-	0.9	0.4	-	0.7	0.4	-	0.3	0.3	-	0.2	0.3	-	0.2	0.2
14	4.1	-	2.5	2.1	-	2.7	1.9	-	1.4	1.7	-	1.0	1.0	-	1.1	0.9	-	0.5	0.5	-	0.7	0.4	-	0.3	0.4	-	0.2	0.2	-	0.2
15	2.7	2.9	4.8	2.3	2.8	4.4	2.2	1.5	2.7	1.5	1.4	2.0	1.2	1.1	1.9	0.6	0.7	1.2	0.5	0.7	0.9	0.4	0.3	0.6	0.3	0.3	0.5	0.2	0.2	0.4
16	-	1.6	1.5	-	2.2	1.1	-	0.9	1.2	-	0.7	0.7	-	0.9	0.6	-	0.4	0.3	-	0.4	0.3	-	0.2	0.2	-	0.2	0.1	-	0.1	0.1
17	1.7	-	2.2	1.6	-	1.1	1.3	-	0.9	1.1	-	0.7	0.9	-	0.7	0.4	-	0.6	0.3	-	0.3	0.3	-	0.2	0.3	-	0.1	0.2	-	0.1
18	1.5	2.1	2.9	1.2	1.5	1.8	1.5	0.8	1.4	1.1	0.7	1.4	0.4	0.5	1.2	0.3	0.6	0.5	0.3	0.3	0.5	0.3	0.2	0.3	0.2	0.1	0.3	0.1	0.1	0.2
19	-	1.5	1.3	-	1.5	1.2	-	0.7	0.8	-	0.9	0.5	-	0.5	0.5	-	0.3	0.3	-	0.3	0.3	-	0.2	0.1	-	0.2	0.1	-	0.1	0.1
20	2.2	-	1.5	1.3	-	1.2	1.2	-	0.9	1.0	-	0.6	0.7	-	0.7	0.4	-	0.3	0.3	-	0.3	0.3	-	0.2	0.2	-	0.1	0.1	-	0.1
21	1.4	1.2	2.2	1.4	1.5	1.6	0.9	0.8	1.8	0.8	0.6	1.0	0.4	0.7	0.9	0.3	0.3	0.5	0.3	0.3	0.4	0.2	0.2	0.3	0.2	0.1	0.2	0.1	0.1	0.2
22	-	1.0	1.3	-	1.1	0.8	-	0.8	0.4	-	0.4	0.4	-	0.4	0.4	-	0.3	0.3	-	0.3	0.2	-	0.2	0.1	-	0.1	0.1	-	0.1	0.1
23	0.9	-	0.8	0.7	-	0.6	1.0	-	0.4	0.6	-	0.5	0.3	-	0.3	0.2	-	0.2	0.2	-	0.2	0.2	-	0.1	0.1	-	0.1	0.1	-	0.1
24	1.1	0.8	1.3	0.7	0.9	1.4	0.5	0.4	0.8	0.5	0.6	0.5	0.3	0.3	0.6	0.2	0.2	0.3	0.1	0.2	0.3	0.1	0.1	0.1	0.1	0.1	0.1	0.1	0.0	0.1
25	-	1.0	1.1	-	0.9	0.6	-	0.6	0.5	-	0.4	0.4	-	0.4	0.3	-	0.3	0.2	-	0.2	0.2	-	0.1	0.1	-	0.1	0.1	-	0.1	0.1
26	0.6	-	0.6	0.8	-	0.4	0.5	-	0.6	0.4	-	0.2	0.2	-	0.3	0.2	-	0.1	0.2	-	0.1	0.1	-	0.1	0.1	-	0.0	0.1	-	0.1
27	0.7	0.6	1.4	0.4	0.6	0.8	0.4	0.5	0.5	0.4	0.2	0.4	0.3	0.2	0.5	0.1	0.1	0.3	0.1	0.1	0.2	0.1	0.1	0.1	0.1	0.0	0.1	0.1	0.0	0.1
28	-	0.9	0.5	-	0.7	0.4	-	0.3	0.3	-	0.3	0.3	-	0.2	0.2	-	0.2	0.1	-	0.1	0.1	-	0.1	0.1	-	0.1	0.1	-	0.0	0.0
29	0.7	-	0.4	0.4	-	0.6	0.3	-	0.2	0.3	-	0.2	0.2	-	0.2	0.1	-	0.1	0.1	-	0.1	0.1	-	0.0	0.1	-	0.0	0.0	-	0.0
30	0.6	0.6	1.1	0.5	0.7	0.7	0.4	0.3	0.6	0.3	0.3	0.5	0.2	0.2	0.4	0.1	0.1	0.2	0.1	0.1	0.2	0.1	0.1	0.1	0.1	0.1	0.1	0.0	0.0	0.1

Figure 10.
First 30x30 tandem gaps, normalized [%], for the first 150 million primes after 51 million.

depends on the number of distinct residue classes the set $\mathcal{H}$ occupies modulo every prime p.

From the perspective of modular arithmetic, the sets $\mathcal{H}_1$ and $\mathcal{H}_2$ possess the identical divisibility properties. The set of pairwise distances in $\mathcal{H}_1$ is $\{2a, 2b, 2(a+b)\}$, which is the same as in $\mathcal{H}_2$. Since there is no privileged direction on the number line regarding primality constraints, the singular series for both configurations are equal:

$$\mathfrak{S}(\mathcal{H}_1) = \mathfrak{S}(\mathcal{H}_2) \tag{13}$$

Consequently, the asymptotic frequency of the gap pair (a, b) is expected to be equal to that of (b, a), validating the symmetry observed in our experimental data.

3.2.2 Frequency behavior and trends

Let $f(a, b)$ denote the normalized frequency of the ordered tandem gap pair (a, b), where $a, b \in \mathbb{N}$ are encoded gap values. Consistent with the theoretical justification provided in the previous section, our empirical data confirms that:

$$f(a, b) \approx f(b, a), \quad \forall a, b \in \{1, 2, \ldots, 50\} \tag{14}$$

validating that the distribution is symmetric with respect to the diagonal.

Moreover, the frequency function $f(a, b)$ exhibits a characteristic trend with respect to a and b: for fixed b, the function $f(a, b)$ decreases for $a \leq 10$ and then increases slowly for $a > 10$ (and symmetrically in b). That is, we observe:

$$\frac{\partial f(a, b)}{\partial a} < 0 \quad \text{for } a \leq 10, \qquad \frac{\partial f(a, b)}{\partial a} > 0 \quad \text{for } a > 10 \tag{15}$$

and similarly for $\partial f(a, b)/\partial b$.

4. Other perspectives

While this study is empirical in nature, based on large-scale computational data, it is important to discuss how the observed patterns relate to existing theoretical frameworks.

Cramér's model. Cramér's probabilistic model [13] predicts that the gaps between consecutive primes $p_{n+1} - p_n$ should be bounded by $O\left(\left(\log p_n\right)^2\right)$ in the long run. Our empirical data does not directly aim to test such upper bounds, but the observed frequency distributions show a high regularity and symmetry which are qualitatively consistent with the assumption of pseudo-randomness in prime locations that underpins Cramér's model.

Hardy–Littlewood conjectures. The Hardy–Littlewood prime k-tuple conjectures [14] predict the asymptotic frequency of prime constellations, including twin primes. Our results on the relative constancy of twin prime frequency (gap code 1) and its slow decrease with growing intervals align qualitatively with this conjecture, which asserts that such pairs should occur with frequency roughly proportional to $\frac{C_2}{(\log n)^2}$, where C_2 is the twin prime constant.

Empirical contribution. Unlike the aforementioned models, our study focuses on local distributions and structural symmetries in short-range prime gap behavior, such as the frequency of specific gap codes and tandem gap pairs. These aspects are not captured directly by classical asymptotic models and may provide new data-driven directions for refining or testing probabilistic predictions.

On the role of statistical validation. We acknowledge that standard statistical tools such as confidence intervals or goodness-of-fit measures are typically used when working with stochastic or sampled data. However, in our case, the prime numbers are deterministic and globally defined, not subject to random sampling. As such, statistical validation in the conventional sense may not be directly applicable.

That said, descriptive statistical tools can still serve a valuable role in measuring the degree to which empirical distributions (e.g., of prime gaps) resemble theoretical or heuristic models (such as log-linear decay). While this study focuses on structural and frequency-based observations, future work may incorporate statistical modeling to characterize these regularities more precisely, always keeping in mind the fundamentally deterministic nature of the underlying data.

4.1 Future work

The empirical observations presented in this study open several avenues for future theoretical and computational research:

- *Analytical justification of tandem gap patterns.* The observed symmetry and modular constraints in tandem gaps (e.g., the impossibility of certain (a, b) pairs) suggest underlying number-theoretic rules. A rigorous formalization and proof of these constraints could lead to new theorems in additive prime number theory.

- *Quantitative comparison with Cramér-type models.* While our study confirms qualitative aspects of prime pseudo-randomness, future work could focus on extracting quantitative metrics (e.g., empirical variance, autocorrelation) from gap sequences and comparing them with predictions from Cramér's model and its modern refinements.

- *Asymptotic behavior across growing intervals*. Although this study analyzes fixed-size intervals (100M, 150M primes), a natural extension is to track how distributional features evolve as both the size and starting point of the intervals grow—potentially leading to empirically informed conjectures about global asymptotic behavior.

- *Integration into probabilistic models*. The empirical distributions of gap codes and tandem gaps may be used to build or refine stochastic models of prime distribution, possibly as constraints or priors in Cramér-style simulations.

- *Theoretical validation of empirical distribution laws*. Our conjectures suggest that the normalized frequency of gap codes follows an approximately linear decay on a logarithmic scale for small and medium values, and that the set of most frequent gaps remains stable across large intervals. These empirical regularities, if confirmed analytically, could lead to new distribution laws for primes or gap frequencies, independent of but possibly related to classical models such as Cramér's or Hardy–Littlewood's.

  Future research may aim to:

  - Formalize the observed log-linear relationship as a statistical law and study its asymptotic validity;

  - Determine whether the frequency ordering of small gap codes (e.g., 3, 6, 5, 9, 1, 2) remains invariant in the limit as $n \to \infty$;

  - Explore whether these stable distributions can be derived from probabilistic assumptions or sieve-theoretic arguments;

  - Investigate the extent to which such patterns are consistent with known results from analytic number theory, or whether they hint at new structural regularities not yet explained.

5. Conclusions

This study demonstrates the use of a highly optimized prime sieve to generate prime numbers across large contiguous intervals and analyze the resulting prime gaps. Employing an empirical research methodology, we have gained some interesting insights into the distribution of prime numbers.

One key finding is the detailed analysis of prime gaps, revealing distinct patterns that may contribute to the understanding of prime number distribution. The introduction of tandem gaps—defined as the ordered pair of two consecutive gaps between three consecutive primes—has led to a new theorem and a new conjecture, offering a fresh perspective on the statistical properties of prime gaps.

The study also includes a set of conjectures regarding the frequency of prime gaps within sequences of consecutive primes, which, if validated, may impact the theoretical understanding of prime number distributions.

Overall, this study highlights the value of computational techniques in number theory, demonstrating how empirical methodologies can lead to theoretical advancements. The findings offer a modest contribution to the current understanding of prime gaps, potentially opening new directions for future research.

Appendix A

To facilitate independent verification of our results, we encourage researchers to generate their own datasets using custom tools and techniques. This approach minimizes potential bias and avoids the inadvertent propagation of any implementation errors that may exist in our code.

Nevertheless, the entire project used to generate and analyze the data is publicly available as a GitHub repository at https://github.com/mirceag70/PrimeGaps. The repository includes the complete source code for the optimized prime sieve, scripts for creating both the simple and tandem gap datasets, the resulting data in CSV format, and the Excel (XLS) files used for the statistical analysis.

The sieve implementation is based on our improved version of the Sieve of Atkin, described in detail in [9]. On a workstation equipped with an AMD 7900X CPU running at 5 GHz and 64 GB of DDR5-5600 RAM, the generation of both datasets (simple gaps and tandem gaps) takes under 7 seconds.

Disclaimers

Conflict of interest

The authors declare no conflict of interest.

Funding

This research did not receive any specific grant from public, commercial, or not-for-profit funding agencies.

AI usage

This study was entirely conducted and the manuscript originally written without the use of artificial intelligence tools during the first part of 2024. AI was employed only in the final stages of the editorial process, for limited language refinement and formatting assistance, without altering the scientific content, methodology, or interpretations presented.

Author details

Mircea Ghidarcea
University POLITEHNICA of Bucharest—Computer Science, Romania

*Address all correspondence to: mircea.ghidarcea@stud.acs.upb.ro

IntechOpen

© 2025 The Author(s). Licensee IntechOpen. This chapter is distributed under the terms of the Creative Commons Attribution License (http://creativecommons.org/licenses/by/4.0), which permits unrestricted use, distribution, and reproduction in any medium, provided the original work is properly cited.

References

[1] Hadamard J. Sur la distribution des zéros de la fonction $\zeta(s)$ et ses conséquences arithmétiques. Bulletin de la Société Mathématique de France. 1896;**24**:199-220. DOI: 10.24033/bsmf.545

[2] de la Vallee Poussin C-J. Recherches analytiques sur la theorie des nombres premiers. Annales de la Societe Scientifique de Bruxelles. 1896;**20**: 183-256. Available from: http://catalog.hathitrust.org/Record/100413897 [Accessed: July 4, 2025]

[3] Erdos P. The difference of consecutive primes. Duke Mathematical Journal. 1940;**6**(2):438-441. DOI: 10.1215/S0012-7094-40-00635-4

[4] Selberg A. An elementary proof of the prime-number theorem. Annals of Mathematics, Second Series. 1949;**50**(2): 305-313. Available from: https://www.math.lsu.edu/~mahlburg/teaching/handouts/2014-7230/Selberg-ElemPNT1949.pdf [Accessed: July 4, 2025]

[5] Zhang Y. Bounded gaps between primes. Annals of Mathematics. 2014; **179**:1121-1174. Available from: https://annals.math.princeton.edu/2014/179-3/p07 [Accessed: July 4, 2025]

[6] Maynard J. Small gaps between primes. Annals of Mathematics. 2015; **181**:383-413. Available from: https://annals.math.princeton.edu/wp-content/uploads/annals-v181-n1-p07-p.pdf [Accessed: July 4, 2025]

[7] Ghidarcea M, Popescu D. Prime Number Sieving—A Systematic Review with Performance Analysis. Algorithms. 2024;**17**:157. DOI: 10.3390/a17040157

[8] Ghidarcea M, Popescu D. Prime Time Tactics—Sieve Tweaks and Boosters. Algorithms. 2024;**17**:291. DOI: 10.3390/a17070291

[9] Ghidarcea M, Popescu D. SIEVE OF ATKIN REVISITED. The Scientific Bulletin, University POLITEHNICA of Bucharest. 2024;**86**(2):15-26. WOS: 001247513600002. Available from: https://www.scientificbulletin.upb.ro/rev_docs_arhiva/fullb53_863029.pdf [Accessed: July 4, 2025]

[10] Ghidarcea M, Popescu D. Static Wheels in Fast Sieves. Journal of Control Engineering and Applied Informatics. 2024;**26**:36-43. DOI: 10.61416/ceai.v26i1.8860

[11] Ghidarcea M, Popescu D. GPGPU and Sundaram. The Scientific Bulletin, University POLITEHNICA of Bucharest. 2024;**86**:15-26. WOS: 001380943300003. Available from: https://www.scientificbulletin.upb.ro/rev_docs_arhiva/fullb1a_720372.pdf [Accessed: July 4, 2025]

[12] Ghidarcea M, Popescu D. Clasic Linear Sieved Revisited. Journal of Control Engineering and Applied Informatics. 2024;**26**:25-36. DOI: 10.61416/ceai.v26i4.9098

[13] Cramér H. On the order of magnitude of the difference between consecutive prime numbers. Acta Arithmetica. 1936;**2**(1):23-46. Available from: http://eudml.org/doc/205441 [Accessed: July 6, 2025]

[14] Hardy GH, Littlewood JE. Some problems of 'Partitio Numerorum': III. On the expression of a number as a sum of primes. Acta Mathematica. 1923;**44**: 1-70. DOI: 10.1007/BF02403921

Chapter 3

Matula Numbers and their Applications

Audace A.V. Dossou-Olory and Sodiq A. Mojeed

Abstract

This chapter focuses on a bijective function between the set of rooted trees and the set of positive integers. It emphasizes some applications of this bijective map in chemistry and combinatorics.

Keywords: Matula numbers, rooted trees, bijection, natural numbers, combinatorics

1. Introduction

In 1968, American mathematician David Matula defined an explicit bijection between the sets of rooted trees (up to isomorphism) and positive natural numbers [1]. The same bijection was obtained in an independent study [2] by F. Göbel in 1980. Hence, it is sometimes called the Matula-Göbel number. The correspondence assigns to each rooted tree T (up to isomorphism) a natural number called the Matula number of T. Henceforth, in this chapter, we shall refer to the set of positive integers denoted $\mathbb{N}^*$ as the set of natural numbers, where $\mathbb{N}^* = \mathbb{N}\backslash\{0\}$.

1.1 Definition

Let T be a rooted tree with root u and p_t be the t-th prime number. That is, $p_1 = 2, p_2 = 3, p_3 = 5$, etc. The number $n(T,u)$, called the Matula number of T, can be obtained recursively as follows: if (T,u) is the one-vertex tree, then $n(T,u) = 1$. However, if $(T_1,v_1), (T_2,v_2), \ldots, (T_d,v_d)$ are all the branches of (T,u), then

$$n(T,u) = \prod_{i=1}^{d} p_{t_i}, \quad d \geq 1, \quad \text{where } t_i = n(T_i,v_i). \tag{1}$$

Moreover, if (T_i,v_i) is the one-vertex tree for any $i \in \{1,2,3, \ldots, d\}$, then $n(T_i,v_i) = 1$. Otherwise, $n(T_i,v_i)$ is obtained recursively according to Eq. (1). Conversely, given a natural number $n \in \mathbb{N}^*$, we can reconstruct its associated rooted tree (up to isomorphism) by following the recursive procedure in the reversed order.

1.1.1 Example

This example illustrates how to calculate the Matula number of the rooted tree in **Figure 1** using the recursive multiplication procedure described in Definition 1.1.

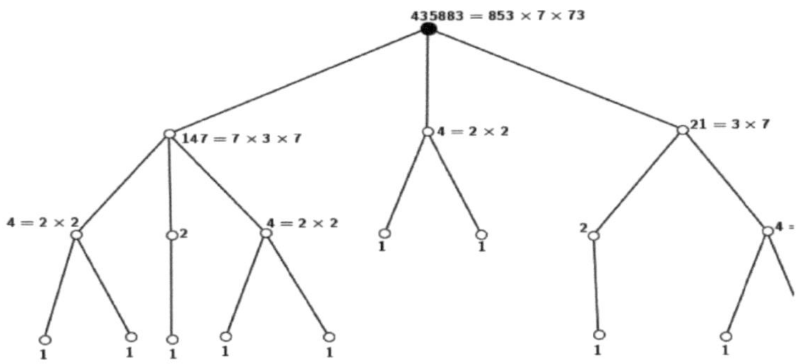

Figure 1.
The Matula number of a rooted tree.

The Matula number of this tree is the product of the prime numbers indexed by all its branches. Moreover, we know that each leaf in the tree corresponds to the natural number 1. We first calculate the Matula numbers of all the branches, and then we determine the Matula number of the tree itself. The last step gives us

$$\begin{aligned} p_{147} \times p_4 \times p_{21} &= 853 \times 7 \times 73 \\ &= 435883. \end{aligned} \tag{2}$$

Furthermore, we provide proof for a proposition that claims that each natural number has a unique rooted tree associated with it [2].

1.1.2 Proposition

For each natural number k, there exists a unique rooted tree (T, u) such that $n(T, u) = k$.

Proof: Suppose by contradiction, k is the smallest natural number for which there does not occur an associated rooted tree (T, u), such that $n(T, u) = k$. If k is prime, then there exists a positive integer t, such that $p_t = k$. Since $p_t = k$, we have $t < k$. Thus, by the hypothesis on k, there is a rooted tree (T', u') whose Matula number is t. If we attach a pendent edge (a new vertex adjacent to the root only), say e, to the root of (T', u'), and we choose the other endpoint of e as our new root, we would obtain the rooted tree (T, u). This is because $n(T, u) = p_t = k$, where $t = n(T', u')$, which is a contradiction.

If k is composite, then we can express k as a product of primes say, $k = k_1 \cdot k_2 \cdot \dots \cdot k_p$ (with multiplicity where possible). Then, as written above, there exists a rooted tree for each of the prime numbers in this product. Thus, we can obtain a new rooted tree (T, u) by identifying the roots of each of the individual rooted trees having Matula numbers $k_1, k_2, \dots, k_p$ with their multiplicities. Thus, the Matula number of the rooted tree (T, u) is k by definition.

To prove uniqueness, let us assume that two different non-isomorphic rooted trees (R, u) and (S, v) have the same Matula number m. Let d_1 be the number of branches in R and d_2 be the number of branches in S. Without loss of generality, we assume m is composite and $d_1 \leq d_2$. Since R is different from S, then there is at least a branch, say, S_1 of S which is not a branch of R (since otherwise, all the branches of S are also the branches of R). As a result, the contribution of S_1 (a prime number) to the Matula number of S will be different from the individual contributions of all the branches of R when obtaining the Matula number of R. Thus, since every natural number greater than 1 admits a unique prime decomposition, we obtain $n(R,u) \neq n(S,v)$, which is a contradiction. Therefore, we conclude that there exists a unique rooted tree corresponding to each positive integer.

2. Some applications of Matula numbers

In the past three decades, Matula numbers have been used in various fields of sciences, particularly in chemistry, analytic combinatorics, coding theory, and quantum mechanics. In this session, we briefly describe their applications in two of these fields.

2.1 Applications of Matula numbers in chemistry

David Matula's result was initially perceived as a mere graph-theoretical curious work that did not have any real-life application until Seymour Elk showed the possibility of using Matula numbers to encode the chemical nomenclatures of hydrogen-suppressed alkanes [3]. Seymour Elk also showed that the Matula-type representation is applicable for assigning a unique name to all possible combinations of either polybenzene or polymantane modules [4].

In organic chemistry, alkanes are known as hydrocarbons with single-chain structures. They have a general molecular formula C_nH_{2n+2}, where $n = 1,2,3,$ etc., with a difference of $-CH_2$ between successive members in their homogeneous group. Their structures exhibit the property that ensures the existence of a single covalent bond between successive carbon atoms [5]. In addition, each carbon has a valency of 4 as shown in **Figure 2**. However, in the hydrogen-suppressed structure, only the carbon atoms are represented; each interior and exterior carbon atom has a valency of 2 and 1, respectively, see **Figure 3**.

Moreover, we can view the structure of each alkane as a connected acyclic graph to allow us to use Matula-type nomenclature to encode them. Let n be the number of carbon atoms in an alkane compound. We can view the carbon atoms in any n-vertex hydrogen-suppressed alkane as the nodes in an n-vertex path graph. If $n = 1$, we shall have the one-vertex tree representing the hydrogen-suppressed methane having

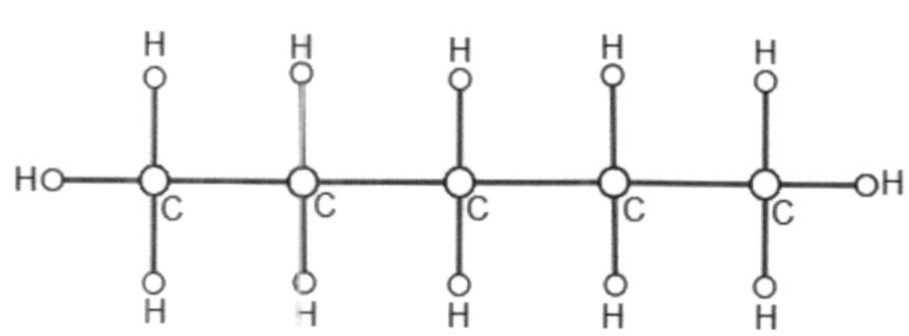

Figure 2.
Propane.

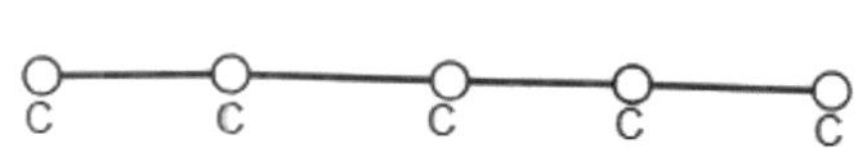

Figure 3.
Hydrogen-suppressed propane.

Matula number 1. However, if $n > 1$, there are n possible nodes that can be designated as the root of such an alkane. In this case, there will be a different Matula number for each choice of the carbon atom selected as the root. For this reason, Seymour Elk suggested that the lowest Matula number should be used as the canonical name of such an alkane [4, 6]. Using this type of single-integer nomenclature helps chemists improve the process of coding, storing, and retrieving information about alkanes in computerized chemical information systems [7].

2.2 Applications of Matula numbers in analytic combinatorics

Analytic combinatorics focuses on enabling precise quantitative predictions of the properties of large combinatorial structures using some generating functions, complex analysis, and integral transforms [8]. In the last three decades, several graph theorists have formulated some results that enable mathematicians to precisely determine some properties of rooted trees directly from their Matula numbers. We now define some statistics of rooted trees and discuss the mathematical frameworks underpinning the formulas for obtaining them from the Matula numbers of rooted trees. These formulas help graph theorists circumvent the time-consuming approach of constructing large-vertex rooted trees before they can determine some of their structural properties.

2.2.1 Definition

Let (T, u) be a rooted tree. The order of T is the number of vertices of T; the width of T is the number of its leaves. The distance between two vertices a and b of T, denoted by $d_T(a, b)$, is the number of edges along the path from a to b. Its diameter is the length of a longest path in T. The height of T can be defined as the length of the longest path from the root. The eccentricity of any vertex v of T, denoted by $E(v)$, is the maximum distance between v and all the vertices of T. Thus,

$$E(v) = \max_{u \in V(T)} d_T(u, v). \tag{3}$$

In **Figure 4**, the distance between vertices l and p is 4, while the distance between vertices p and r is 6. Furthermore, this rooted tree has a diameter and height of 6 and 3, respectively. Moreover, $E(b) = 4$ and $E(h) = 5$.

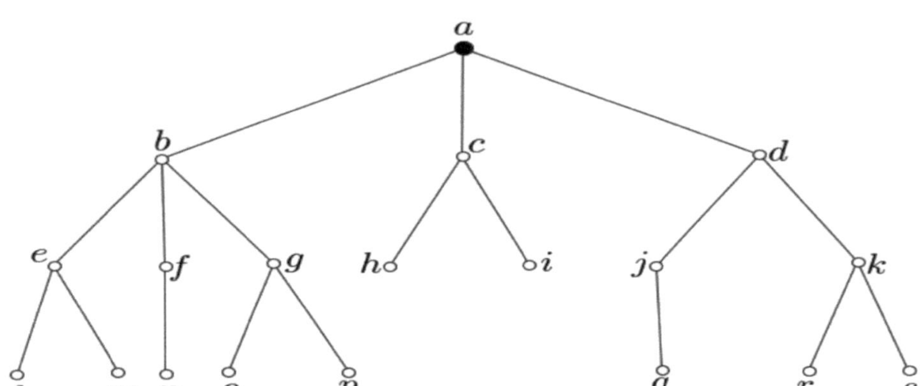

Figure 4.
The topology of a rooted tree.

2.2.2 Mathematical frameworks: Matula numbers vs. tree statistics

Let $\mathcal{T}$ denote the sets of all rooted trees. We define a map $n : \mathcal{T} \to \mathbb{N}^*$ where n maps each rooted tree in $\mathcal{T}$ to its corresponding Matula number. Hence, $n(T, u) = 1$ if T is the one-vertex tree in $\mathcal{T}$. Furthermore, let τ be the inverse map of n, that is, $\tau : \mathbb{N}^* \to \mathcal{T}$, where $\tau(1)$ corresponds to the one-vertex tree in $\mathcal{T}$. The maps τ and n are well defined based on Proposition 1.1.2 and the formal definition of Matula numbers given in Definition 1.1.

Suppose $n \in \mathbb{N}^*$ is the Matula number of a rooted tree T, then $\tau(n) = T$ (see **Figure 5**). If n is a prime number, then there exists t such that $t < p_t = n$. Then, $\tau(p_t)$ is formed by attaching a pendent edge to $\tau(t)$ as shown in **Figure 6**. This implies that $\tau(p_t)$ has only one branch which is $\tau(t)$. However, if n is composite, then we can write $n = ab$, for some $a, b \geq 2$. In this case, we take a and b as the Matula numbers of two rooted trees $\tau(a)$ and $\tau(b)$ identifying their roots to form $\tau(ab)$ as illustrated in **Figure 7**.

In what follows, we describe how we use the notations representing different statistics of rooted trees in this chapter.

Let X be a statistic of rooted trees. Suppose a rooted tree T has Matula number n, we denote the value of X for T by $X(n)$. In addition, $\Omega(n)$ is the number of prime factors of n (with multiplicity).

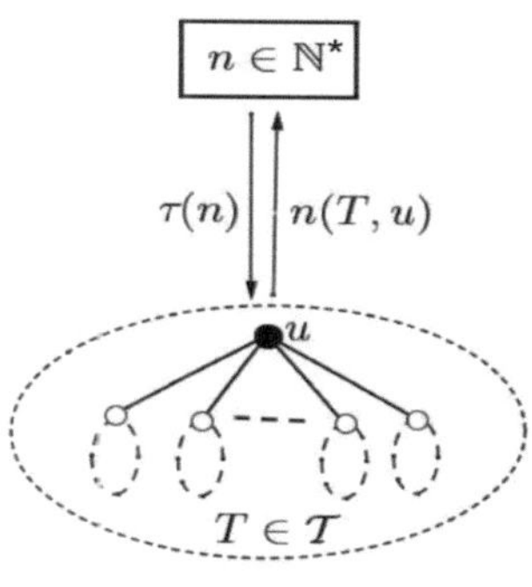

Figure 5.
A bijection between the sets of rooted trees and natural numbers.

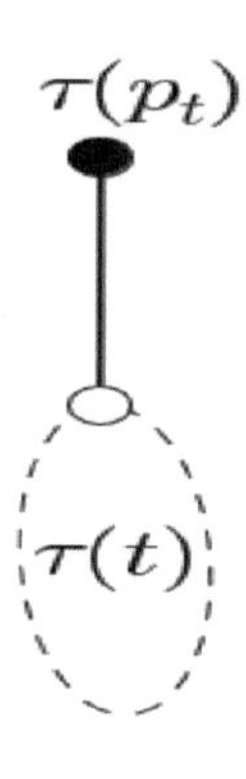

Figure 6.
An rooted tree extended by a pendant edge.

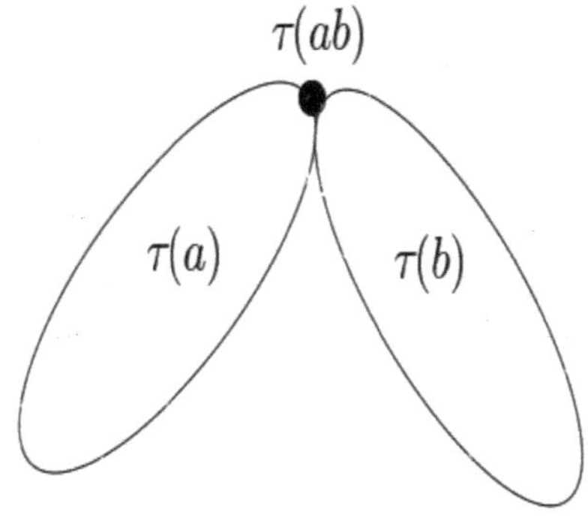

Figure 7.
A rooted tree formed by identifying the roots of $\tau(a)$ and $\tau(b)$.

2.2.3 Statistics of rooted trees from their Matula numbers

The following formulas can be used to determine some statistics of rooted trees from their associated Matula numbers. Many such results have been studied and proven in literature [3, 7, 9]. Moreover, we will provide a few of them together with their proofs.

Let n be the Matula number of a rooted tree $\tau(n)$.

2.2.4 Proposition

Let $V(n)$ denote the order of $\tau(n)$. Then,

$$V(n) = \begin{cases} 1, & \text{if} \quad n = 1; \\ 1 + V(t), & \text{if} \quad n = p_t; \\ V(a) + V(b) - 1, & \text{if} \quad n = ab, a, b \geq 2. \end{cases} \tag{4}$$

Proof: For the one-vertex tree, $V(n) = 1$. However, if $n = p_t$ is a prime number, then by **Figure 6**, the number of vertices in $\tau(p_t)$ is 1 more than the number of vertices in $\tau(t)$; hence, $V(p_t) = 1 + V(t)$. Moreover, if n is composite, then by **Figure 7**, the number of vertices in $\tau(ab)$ is 1 less than the sum of the number of vertices in $\tau(a)$ and $\tau(b)$. Hence, $V(ab) = V(a) + V(b) - 1$. It is necessary to subtract 1 from the sum to avoid double counting the roots of $\tau(a)$ and $\tau(b)$.

2.2.5 Proposition

Let $L(n)$ denote the number of leaves of $\tau(n)$. Then,

$$L(n) = \begin{cases} 0, & \text{if} \quad n = 1; \\ 1, & \text{if} \quad n = 2; \\ L(t), & \text{if} \quad n = p_t, t \geq 2; \\ L(a) + L(b), & \text{if} \quad n = ab, a, b \geq 2. \end{cases} \tag{5}$$

Proof: If $n = 1$, it is sufficient to say that the one-vertex tree has no leaf; hence, $L(n) = 0$. However, if $n = 2$, the associated tree $\tau(n)$ is the path of length 1 (with the root at one of its endpoints). In this case, $\tau(n)$ has only one leaf. Therefore, $L(n) = 1$. Furthermore, if $n = p_t$ is a prime number greater than 2, then by **Figure 6**, all the leaves of $\tau(p_t)$ are precisely the leaves of $\tau(t)$; hence, $L(p_t) = L(t)$. Finally, if n is composite, then by **Figure** 7, the number of leaves of $\tau(ab)$ is the sum of the number of leaves of $\tau(a)$ and $\tau(b)$. Therefore, $L(ab) = L(a) + L(b)$.

Note that, for rooted trees, there is a difference between leaves and pendent vertices. A leaf in a rooted tree is a *non-root* vertex of degree 1, while a pendent vertex is a vertex of degree 1 (including the root if it has degree 1). We encourage our readers to observe the difference between the results in Propositions 2.2.5 and 2.2.6.

2.2.6 Proposition

Let $PV(n)$ denote the number of pendent vertices of $\tau(n)$. Then,

$$PV(n) = \begin{cases} 0, & \text{if} \quad n = 1; \\ 2, & \text{if} \quad n = 2; \\ 1 + L(t), & \text{if} \quad n = p_t, t \geq 2; \\ L(a) + L(b), & \text{if} \quad n = ab, a, b \geq 2. \end{cases} \tag{6}$$

Proof: Since the one-vertex tree has no pendent vertex, $PV(n) = 0$. Moreover, if $n = 2$, the associated tree $\tau(n)$ has two pendent vertices because its root and leaf have degree 1. However, if $n = p_t$ is a prime number greater than 2, then by **Figure 6**, the number of pendent vertices of $\tau(p_t)$ is 1 more than the number of leaves of $\tau(t)$, where 1 accounts for the contribution of the root of $\tau(p_t)$. Hence, $PV(p_t) = 1 + L(t)$. Finally, if n is composite, then by **Figure** 7, the number of pendent vertices of $\tau(ab)$ is the sum of the number of leaves of $\tau(a)$ and $\tau(b)$. Thus, $PV(ab) = L(a) + L(b)$.

2.2.7 Proposition

Let $E(n)$ denote the number of edges of $\tau(n)$. Then,

$$E(n) = \begin{cases} 0, & \text{if} \quad n = 1; \\ 1 + E(t), & \text{if} \quad n = p_t; \\ E(a) + E(b), & \text{if} \quad n = ab, a, b \geq 2. \end{cases} \tag{7}$$

Proof: If $n = 1$, it is sufficient to say that the one-vertex tree has no edge; hence, $E(n) = 0$. Moreover, if $n = p_t$ is a prime number, then by **Figure 6**, the number of edges of $\tau(p_t)$ is 1 more than the number of edges of $\tau(t)$; hence, $E(p_t) = 1 + E(t)$. However, if n is composite, then by **Figure** 7, the number of edges of $\tau(ab)$ is the sum of the number of edges of $\tau(a)$ and $\tau(b)$. As a result, $E(ab) = E(a) + E(b)$..

We recall that the height of T is the maximum distance from the root of T to a leaf of the tree. Furthermore, the eccentricity of the root of T is the maximum distance between the root and all the vertices of T. Hence, the height of T is precisely the eccentricity of its root. Hence, it is sufficient to present the result in Proposition 2.2.8 for the height of any rooted tree T.

2.2.8 Proposition

Let $H(n)$ denote the height of $\tau(n)$. Then,

$$H(n) = \begin{cases} 0, & \text{if} \quad n = 1; \\ 1 + H(t), & \text{if} \quad n = p_t; \\ \max(H(a), H(b)), & \text{if} \quad n = ab, a, b \geq 2. \end{cases} \tag{8}$$

Proof: If $n = 1$, then $\tau(n)$ is the one-vertex tree; hence, $H(n) = 0$. Moreover, if $n = p_t$ is a prime number, then by **Figure 6**, the height of $\tau(p_t)$ is 1 more than the height of $\tau(t)$. Therefore, $H(p_t) = 1 + H(t)$. Finally, if n is composite, then by **Figure 7**, the height of $\tau(ab)$ is the maximum of the heights of $\tau(a)$ and $\tau(b)$. Hence, $H(ab) = \max(H(a), H(b))$..

2.2.9 Proposition

Let $\Delta(n)$ denote the maximum vertex degree of $\tau(n)$. Then,

$$\Delta(n) = \begin{cases} 0, & \text{if} \quad n = 1; \\ \max(\Delta(t), 1 + \Omega(t)), & \text{if} \quad n = p_t; \\ \max(\Delta(a), \Delta(b), \Omega(a) + \Omega(b)), & \text{if} \quad n = ab, a, b \geq 2, \end{cases} \tag{9}$$

where $\Omega(t)$ is the number of prime factors of t (with multiplicity).

Proof: It is obvious that the one-vertex tree has degree 0; hence, if $n = 1$, $\Delta(n) = 0$. However, if $n = p_t$ is a prime number, then $\tau(p_t)$ has only one branch which is $\tau(t)$. In this case, one can observe two possibilities: either a vertex with the maximum degree is a root-vertex or a non-root vertex of $\tau(t)$. For the former, we determine $1 + \Omega(t)$, whereas for the latter, we find $\Delta(t)$. Hence, $\Delta(p_t) = \max(\Delta(t), 1 + \Omega(t))$..

If n is composite, then we can write $n = ab$, for some $a, b \geq 2$, where a and b are the Matula numbers of two rooted trees $\tau(a)$ and $\tau(b)$. In this case, we observe three possibilities:

1. The maximum vertex degree is the degree of a non-root vertex of $\tau(a)$.

2. The maximum vertex degree is the degree of a non-root vertex of $\tau(b)$.

3. The maximum vertex degree is the degree of the root of $\tau(ab)$.

Therefore, $\Delta(ab) = \max(\Delta(a), \Delta(b), \Omega(a) + \Omega(b))$.

3. Conclusion

In this chapter, we discussed David Matula's explicit bijection between the sets of rooted trees and natural numbers. This correspondence reveals that, to each positive integer, there is a unique associated rooted tree and vice versa. Therefore, one can easily encode rooted trees using natural numbers. This encoding characteristic of rooted trees is an integral part of the applications of Matula numbers in organic chemistry and analytic combinatorics.

Author details

Audace A.V. Dossou-Olory[1*] and Sodiq A. Mojeed[2]

1 University of Abomey-Calavi, Benin

2 African Institute for Mathematical Science, Ghana

*Address all correspondence to: audace@aims.ac.za

IntechOpen

© 2025 The Author(s). Licensee IntechOpen. This chapter is distributed under the terms of the Creative Commons Attribution License (http://creativecommons.org/licenses/by/4.0), which permits unrestricted use, distribution, and reproduction in any medium, provided the original work is properly cited.

References

[1] Matula DW. A natural rooted tree enumeration by prime factorization. SIAM Review. 1968;**10**(2):273

[2] Göbel F. On a 1-1 correspondence between rooted trees and natural numbers. Journal of Combinatorial Theory, Series B. 1980;**29**(1):141-143

[3] Gutman I, Yeh YN. Deducing properties of trees from their Matula numbers. Publications de L'Institut Mathématique. 1993;**53**(67):17-22

[4] Elk SB. A canonical ordering of polybenzenes and polymantanes using a prime number factorization technique. Journal of Mathematical Chemistry. 1990;**4**(1):55-68

[5] Bahl B, Bahl A. A Textbook of Organic Chemistry. New Delhi: S. Chand & Co. Ltd.; 2017

[6] Elk SB, Gutman I. Further properties derivable from the Matula number of an alkane. Journal of Chemical Information and Computer Sciences. 1994;**34**: 331-334. Available from: https://api.semanticscholar.org/CorpusID:39186873

[7] Gutman I, Ivić A. On Matula Numbers. Discrete Mathematics. 1996; **150**(1-3):131-142

[8] Flajolet P, Sedgewick R. Analytic Combinatorics. New York: Cambridge University Press; 2019

[9] Deutsch E. Rooted tree statistics from Matula numbers. Discrete Applied Mathematics. 2012;**160**(15):2314-2322

Section 2

Computational, Applied, and Algorithmic Perspectives

Chapter 4

On Efficient Single-Step Scheme for Solving p-Adic Polynomial in $\mathbb{Z}_\mathbf{p}$: Convergence and Fractal Analysis

Mudassir Shams and Bruno Carpentieri

Abstract

We propose a new single-step iterative method for solving nonlinear equations in the ring of p-adic integers $\mathbb{Z}_p$. The method is derived using a p-adic Taylor expansion, and its convergence is rigorously established in ultrametric spaces under Lipschitz-type conditions. Furthermore, the dynamical behavior of the scheme is explored by generating the basins of attraction over discretized $\mathbb{Z}_p$, which reveals fractal structures influenced by the choice of the prime p and the initial approximation. Numerical experiments in $\mathbb{Z}_3$ and $\mathbb{Z}_5$ confirm the theoretical results and demonstrate that the scheme is stable, sensitive to initial values, and computationally efficient. The observed p-adic fractal patterns resemble classical complex fractals, offering new insights into root-finding dynamics in non-Archimedean fields.

Keywords: p-adic numbers, iterative methods, $\mathbb{Z}_p$, convergence analysis, fractals, polynomial equations

1. Introduction

Kurt Hensel introduced the concept of p-adic numbers in 1897, marking a major development in number theory [1]. The field $\mathbb{Q}_\mathbf{p}$ is constructed as the completion of $\mathbb{Q}$ with respect to the **p**-adic norm, and this process differs fundamentally from the construction of the real numbers $\mathbb{R}$. For many years, **p**-adic numbers were regarded primarily as abstract objects with little relevance beyond pure mathematics. However, starting from the 1960s, their importance began to extend well beyond theory, finding applications in diverse areas of physics and mathematics, including string theory, quantum mechanics, quantum cosmology, and dynamical systems [2]. More recently, **p**-adic methods have also emerged in biological and computational settings—particularly in bioinformatics—where they are used to model hierarchical structures and discrete processes [3]. For an in-depth introduction to the theory of p-adic numbers, we refer the reader to Koblitz [4], while Razikov [5] provides a concise overview of their applications across mathematics, physics, and biology.

A classical application of $\mathbf{p}$-adic numbers is given by Hensel's lemma, which guarantees the existence and uniqueness of polynomial roots modulo powers of a prime $\mathbf{p}$. This result serves as the theoretical foundation for iterative root-finding methods in $\mathbb{Z}_{\mathbf{p}}$. Building on this idea, several authors have investigated the computation of q-th roots of $\mathbf{p}$-adic numbers in $\mathbb{Q}_{\mathbf{p}}$, where q is a prime smaller than $\mathbf{p}$ [6]. The works of Kecies and Zerzaihi [6–8] further extended these concepts to numerical schemes for $\mathbf{p}$-adic root-finding. Related problems—such as determining multiplicative inverses modulo powers of a prime—have been studied by Dumas [9] and Knapp and Xenophontos [10]. Together, these contributions underscore both the theoretical depth and the computational complexity associated with $\mathbf{p}$-adic arithmetic.

Despite these advancements, most existing studies address only specific cases of $\mathbf{p}$-adic root finding, such as the computation of square or cubic roots, and do not extend to general higher-order polynomial equations with intermediate terms in $\mathbb{Z}_{\mathbf{p}}[x]$. Moreover, while classical numerical analysis typically focuses on fixed-precision arithmetic and counts of function evaluations, $\mathbf{p}$-adic numerical analysis demands a different perspective. In particular, arithmetic operations modulo $\mathbf{p}^k$ often involve multiple machine-level instructions, and modular inversions are computationally expensive. As a result, minimizing the number of iterations alone is not sufficient; the precision and size of intermediate computations must also be taken into account to optimize the overall computational cost. Recent studies have begun to investigate these aspects more systematically, showing that the lengths of intermediate results frequently follow simple linear recurrence relations, which in turn allow the total asymptotic cost to be derived by summing geometric series [11].

Classical iterative methods such as the Newton–Raphson, secant, and fixed-point schemes have been adapted to the $\mathbf{p}$-adic setting to approximate roots of polynomials in $\mathbb{Z}_{\mathbf{p}}$. For instance, Knapp and Xenophontos [12] demonstrated the computation of modular inverses using classical root-finding techniques, while Zerzaihi et al. [13, 14] applied fixed-point, secant, and Newton methods to determine square and cubic roots of $\mathbf{p}$-adic numbers. Ignacio et al. [15, 16] further refined these approaches by developing algorithms based on the Newton–Raphson iteration for root computation in $\mathbf{p}$-adic fields. However, these contributions primarily focus on low-degree cases and do not consider general higher-order polynomial equations. Moreover, they lack a unified framework that addresses error analysis, convergence visualization, and the optimization of computational cost in the $\mathbf{p}$-adic context.

Motivated by these gaps, this chapter addresses the general problem of solving

$$f(x) = \sum_{k=0}^{n} a_k x^k, \quad \text{with } a_k \in \mathbb{Q}_{\mathbf{p}}. \tag{1}$$

where f is a higher-order polynomial with intermediate terms. We propose a modified $\mathbf{p}$-adic Newton method of the form

$$x^{[t+1]} = x^{[t]} - \frac{f\left(x^{[t]}\right)}{f'\left(x^{[t]}\right) + \alpha f\left(x^{[t]}\right)} \left(\bmod \mathbf{p}^{2^t}\right). \tag{2}$$

where $\alpha \in \mathbb{Z}_{\mathbf{p}}$ is a parameter introduced to improve convergence stability. This scheme generalizes existing $\mathbf{p}$-adic root-finding methods and offers a practical computational framework for lifting roots from $\mathbf{p}$ to $\mathbf{p}^k$. Alongside the theoretical convergence analysis, we examine the computational aspects such as the number of iterations, CPU time, memory usage, and error propagation. Moreover, we provide visualizations of the lifting process and convergence behavior, which offers an intuitive interpretation of **p**-adic quadratic convergence based on the Taylor expansion in the **p**-adic setting.

The main contributions of this work can be summarized as follows:

1. We develop a modified **p**-adic numerical algorithm for efficiently computing the zeros of general higher-order polynomials in $\mathbb{Z}_{\mathbf{p}}[x]$.

2. Using **p**-adic norm arguments and a Taylor expansion approach, we provide a rigorous local convergence analysis, establishing precise criteria for convergence and demonstrating quadratic error decay.

3. We conduct a detailed computational study of intermediate quantities, highlighting the relationship between iteration counts, arithmetic cost, and accuracy requirements for nonlinear equations.

4. The iterative lifting process, error reduction, fractal behavior, and convergence dynamics of the proposed scheme are illustrated using diagrams and numerical test problems.

5. We show that the scheme can be applied to a broader class of polynomial equations arising in science, medicine, economics, and engineering, thereby extending and generalizing previous approaches.

This study is distinct in that it combines theoretical **p**-adic analysis with the practical numerical implementation of **p**-adic iterative techniques, including error control and assessments of computational efficiency. It provides a unified framework for a general family of **p**-adic polynomial root-finding methods that is both computationally feasible and theoretically rigorous. This contrasts with earlier works, which typically focuses either solely on theoretical aspects or on specific root-finding cases without offering a broader, systematic approach.

The remainder of this chapter is organized as follows. Section 2 reviews the essential background on **p**-adic numbers, Hensel's lemma, and classical **p**-adic root-finding methods, and introduces the proposed family of iterative schemes in the **p**-adic setting. Section 3 presents the numerical experiments that illustrate convergence behavior, computational cost, fractal analysis, and residual error dynamics for representative polynomial examples. Finally, Section 4 summarizes the main results and outlines possible directions for future research in **p**-adic computational methods for complex problems arising in science and engineering.

2. Preliminaries on $\mathbb{Q}_{\mathbf{p}}$

In this section, we recall the fundamental definitions and properties of the field of **p**-adic numbers $\mathbb{Q}_{\mathbf{p}}$ [17–19], which are essential for understanding the numerical schemes developed in this study.

Definition 1. [20] Let $\mathbb{Z}$ and $\mathbb{Q}$ denote the sets of integers and rational numbers, respectively. For a fixed prime number **p** and any integer $x \in \mathbb{Z} \backslash \{0\}$, the **p**-*adic valuation* $v_{\mathbf{p}}(x)$ is defined as the highest power of **p** dividing x, i.e.,

$$v_{\mathbf{p}}(x) = \text{the largest integer } k \text{ such that } \mathbf{p}^k | x. \tag{3}$$

This definition extends multiplicatively to all nonzero rational numbers $x = \frac{a}{b} \in \mathbb{Q}$ by

$$v_{\mathbf{p}}\left(\frac{a}{b}\right) = v_{\mathbf{p}}(a) - v_{\mathbf{p}}(b). \tag{4}$$

The corresponding **p**-*adic norm* (or absolute value) on $\mathbb{Q}$ is given by

$$|x|_{\mathbf{p}} = \mathbf{p}^{-v_{\mathbf{p}}(x)}. \tag{5}$$

This norm satisfies the ultrametric (non-Archimedean) property

$$|x+y|_{\mathbf{p}} \leq \max\left\{|x|_{\mathbf{p}}, |y|_{\mathbf{p}}\right\},$$

which distinguishes **p**-adic analysis from its real counterpart. The completion of $\mathbb{Q}$ under the metric induced by $|\cdot|_{\mathbf{p}}$ forms the field of **p**-adic numbers $\mathbb{Q}_{\mathbf{p}}$.

3. Basic definitions and properties of the p-adic norm

Definition 2. [20] Let **p** be a fixed prime number. The **p**-*adic norm*

$$|\cdot|_{\mathbf{p}} : \mathbb{Q} \rightarrow \{\mathbf{p}^n : n \in \mathbb{Z}\} \cup \{0\}$$

is defined for any rational number $x \in \mathbb{Q}$ as

$$|x|_{\mathbf{p}} = \begin{cases} \mathbf{p}^{-v_{\mathbf{p}}(x)}, & \text{if } x \neq 0, \\ 0, & \text{if } x = 0, \end{cases} \tag{6}$$

where $v_{\mathbf{p}}(x)$ denotes the **p**-*adic valuation* of x, defined by

$$v_{\mathbf{p}}(x) = \max\{r \in \mathbb{Z} : \mathbf{p}^r \text{ divides } x\}.$$

In other words, the valuation $v_{\mathbf{p}}(x)$ counts the highest power of **p** dividing x. This definition is extended to all rational numbers $x = \frac{a}{b} \in \mathbb{Q} \backslash \{0\}$, where $a, b \in \mathbb{Z}$ and $b \neq 0$, by

$$v_{\mathbf{p}}\left(\frac{a}{b}\right) = v_{\mathbf{p}}(a) - v_{\mathbf{p}}(b).$$

Hence, the norm of a rational number can be expressed as

$$\left|\frac{a}{b}\right|_{\mathbf{p}} = \mathbf{p}^{-v_{\mathbf{p}}(a)+v_{\mathbf{p}}(b)}.$$

The norm $|\cdot|_{\mathbf{p}}$ induces a metric $d_{\mathbf{p}}$ on the field of rational numbers, defined by

$$d_{\mathbf{p}} : \mathbb{Q} \times \mathbb{Q} \rightarrow \mathbb{R}^+, \qquad d_{\mathbf{p}}(x,y) := |x-y|_{\mathbf{p}}. \tag{7}$$

This metric, called the **p**-*adic metric*, measures the "closeness" of two rational numbers not by their difference in the real sense, but by the divisibility of their difference by powers of **p**. In this sense, two numbers are *closer* in the **p**-adic metric if their difference is divisible by a high power of **p**.

The **p**-adic norm possesses several fundamental properties that distinguish it from the usual absolute value on $\mathbb{R}$ [21]:

1. *Multiplicativity:*

$$|xy|_{\mathbf{p}} = |x|_{\mathbf{p}}|y|_{\mathbf{p}}, \qquad \forall x, y \in \mathbb{Q}.$$

This property ensures that the norm behaves consistently under multiplication.

2. *Ultrametric (Non-Archimedean) Inequality:*

$$|x+y|_{\mathbf{p}} \leq \max\left\{|x|_{\mathbf{p}}, |y|_{\mathbf{p}}\right\},$$

with equality whenever $|x|_{\mathbf{p}} \neq |y|_{\mathbf{p}}$. This inequality implies that in $\mathbb{Q}_{\mathbf{p}}$, every triangle is *isosceles*, and the geometry of the space is totally disconnected.

3. *Division Property:*

$$\left|\frac{x}{y}\right|_{\mathbf{p}} = \frac{|x|_{\mathbf{p}}}{|y|_{\mathbf{p}}}, \qquad \text{for } y \neq 0.$$

Example. Consider $\mathbf{p} = 3$ and $x = 81 = 3^4$. Then

$$v_3(81) = 4, \quad |81|_3 = 3^{-4} = \frac{1}{81}.$$

Similarly, for $x = \frac{9}{27} = \frac{3^2}{3^3}$, we obtain

$$v_3\left(\frac{9}{27}\right) = 2 - 3 = -1, \quad \text{so that} \quad \left|\frac{9}{27}\right|_3 = 3^{-(-1)} = 3.$$

Thus, in the 3-adic sense, $\frac{9}{27}$ has a *larger* norm than 81, reflecting the inverse divisibility structure intrinsic to the **p**-adic world.

Definition 3. [22] The *field of* **p**-*adic numbers*, denoted by $\mathbb{Q}_{\mathbf{p}}$, is the completion of $\mathbb{Q}$ with respect to the metric $d_{\mathbf{p}}$ (or equivalently, the norm $|\cdot|_{\mathbf{p}}$).

For an element $a \in \mathbb{Q}_p$, represented as the limit of a Cauchy sequence $\{a_n\} \subset \mathbb{Q}$, the extended **p**-adic norm is defined by

$$|a|_{\mathbf{p}} = \lim_{n \to \infty} |a_n|_p. \tag{8}$$

This definition is independent of the choice of the representative Cauchy sequence.

3.1 Canonical representation and p-adic integers

Theorem 1 (Canonical Representation of **p**-adic Numbers [23]). *For every element* $a \in \mathbb{Q}_{\mathbf{p}}$, *there exists a unique sequence of integers* $\{a_i\}_{i \geq N}$, *where* $N = v_{\mathbf{p}}(a)$ *and* $0 \leq a_i \leq \mathbf{p} - 1$, *such that*

$$a = a_N \mathbf{p}^N + a_{N+1}\mathbf{p}^{N+1} + a_{N+2}\mathbf{p}^{N+2} + \cdots = \sum_{i=N}^{\infty} a_i \mathbf{p}^i. \tag{9}$$

This series is called the canonical **p**-adic expansion *of a. The coefficients a_i are uniquely determined and form the digits of a in base* **p**. *Unlike the usual real number representation, this expansion extends infinitely to the left rather than to the right, reflecting the non-Archimedean nature of the* **p**-*adic norm.*

Definition 4 (**p**-adic Integers [8]) A **p**-*adic integer* is an element of $\mathbb{Q}_\mathbf{p}$ whose canonical representation (9) involves only nonnegative powers of **p**, i.e., the smallest exponent N is nonnegative. The set of all **p**-adic integers is denoted by $\mathbb{Z}_\mathbf{p}$ and can be characterized equivalently as

$$\mathbb{Z}_\mathbf{p} = \left\{ a \in \mathbb{Q}_\mathbf{p} : |a|_\mathbf{p} \leq 1 \right\}. \tag{10}$$

Hence, every **p**-adic integer can be written uniquely in the form

$$a = a_0 + a_1\mathbf{p} + a_2\mathbf{p}^2 + a_3\mathbf{p}^3 + \cdots = \sum_{i=0}^{\infty} a_i \mathbf{p}^i, \quad 0 \leq a_i \leq \mathbf{p} - 1. \tag{11}$$

This infinite series converges in the **p**-adic norm due to the fact that $\lim_{i\to\infty} |a_i \mathbf{p}^i|_\mathbf{p} = 0$.

The ring $\mathbb{Z}_\mathbf{p}$ is compact and complete, and serves as the ring of integers in the field $\mathbb{Q}_\mathbf{p}$. Moreover, its residue field is $\mathbb{F}_\mathbf{p} = \mathbb{Z}_\mathbf{p}/\mathbf{p}\mathbb{Z}_\mathbf{p}$.

Lemma 2 (Congruence and **p**-adic Distance [24]). *For any $a, b \in \mathbb{Q}_\mathbf{p}$ and integer $m \geq 1$, the following equivalence holds:*

$$a \equiv b \pmod{\mathbf{p}^m} \quad \Leftrightarrow \quad |a - b|_\mathbf{p} \leq \mathbf{p}^{-m}. \tag{12}$$

This relation provides a natural link between modular congruence and the **p**-*adic topology. Two* **p**-*adic numbers are congruent modulo $\mathbf{p}^m$ precisely when their* **p**-*adic distance is not greater than $\mathbf{p}^{-m}$. Consequently, the modular arithmetic over $\mathbb{Z}_\mathbf{p}$ can be interpreted geometrically as neighborhoods in the* **p**-*adic metric space.*

Remark 3. The field $\mathbb{Q}_\mathbf{p}$ is obtained as the completion of $\mathbb{Q}$ with respect to the **p**-adic norm. Every **p**-adic number possesses a unique representation as a convergent series in powers of **p**, analogous to a decimal expansion but governed by the ultrametric convergence. This construction forms the foundation of many number-theoretic algorithms, including Hensel's lemma and **p**-adic lifting methods, which rely on the precision hierarchy induced by the **p**-adic valuation.

3.2 Continuity and differentiability over $\mathbb{Q}_\mathbf{p}$

In classical real analysis, the notions of continuity and differentiability are defined using the Euclidean metric. In contrast, over the field of **p**-adic numbers $\mathbb{Q}_\mathbf{p}$, these concepts are reformulated in terms of the non-Archimedean **p**-adic norm $|\cdot|_\mathbf{p}$, which induces a topology fundamentally different from the Euclidean one. The **p**-adic topology possesses several distinctive features, including the ultrametric inequality and the totally disconnected structure of $\mathbb{Q}_\mathbf{p}$, which result in markedly different behaviors of functions and sequences.

Definition 5 (Continuity [25]): Let $X \subset \mathbb{Q}_{\mathbf{p}}$ be a subset of the field of **p**-adic numbers. A function $f : X \to \mathbb{Q}_{\mathbf{p}}$ is said to be *continuous at a point* $a \in X$ if for every $\varepsilon > 0$, there exists $\delta > 0$ such that

$$|x - a|_{\mathbf{p}} < \delta \Rightarrow |f(x) - f(a)|_{\mathbf{p}} < \varepsilon. \tag{13}$$

The function f is said to be *continuous on* X if it is continuous at every point $a \in X$.

Due to the ultrametric property of the **p**-adic norm, continuity over $\mathbb{Q}_{\mathbf{p}}$ often exhibits stronger local behavior than in $\mathbb{R}$. In particular, for a **p**-adic function f, if $|f(x) - f(a)|_{\mathbf{p}}$ is small for one point x within a small neighborhood of a, it typically remains equally small for all points within that neighborhood, reflecting the "clopen" (simultaneously closed and open) nature of **p**-adic balls.

Definition 6 (Differentiability [26]): Let $X \subset \mathbb{Q}_{\mathbf{p}}$ and let $a \in X$ be an accumulation point. The function $f : X \to \mathbb{Q}_{\mathbf{p}}$ is said to be *differentiable at* a if the limit

$$f'(a) = \lim_{x \to a} \frac{f(x) - f(a)}{x - a} \tag{14}$$

exists with respect to the **p**-adic norm $|\cdot|_{\mathbf{p}}$. The value $f'(a)$, if it exists, is called the **p**-*adic derivative* of f at a.

Differentiability in $\mathbb{Q}_p$ inherits several familiar properties from real analysis but differs in others. For instance, if $f'(a) = 0$, then the function f is locally constant on some **p**-adic neighborhood of a, which does not occur in the real setting. Moreover, analytic functions in $\mathbb{Q}_{\mathbf{p}}$ are often defined through convergent power series whose radii of convergence depend on the p-adic absolute value of the coefficients.

Remark 4. Every polynomial function $f \in \mathbb{Q}_{\mathbf{p}}[x]$ is continuous and differentiable on $\mathbb{Q}_{\mathbf{p}}$. Furthermore, the derivative is obtained by the formal differentiation rule

$$f'(x) = \sum_{k=1}^{n} k a_k x^{k-1},$$

where $f(x) = \sum_{k=0}^{n} a_k x^k$ with coefficients $a_k \in \mathbb{Q}_{\mathbf{p}}$. This property extends to rational functions defined wherever the denominator is nonzero in $\mathbb{Q}_{\mathbf{p}}$.

3.3 Hensel's lemma and root lifting

One of the most fundamental results linking continuity, differentiability, and the solvability of polynomial equations in the **p**-adic setting is *Hensel's Lemma*. It serves as the **p**-adic analogue of Newton's method and provides a constructive criterion for lifting modular solutions to genuine the **p**-adic roots.

Theorem 5 (Hensel's Lemma [27]). *Let* $f(x) = C_0 + C_1 x + \cdots + C_q x^q \in \mathbb{Z}_{\mathbf{p}}[x]$ *and let* $f'(x)$ *denote its formal derivative. Suppose there exists* $a_0 \in \mathbb{Z}_{\mathbf{p}}$ *such that*

$$f(a_0) \equiv 0 \pmod{\mathbf{p}}, \quad f'(a_0) \not\equiv 0 \pmod{\mathbf{p}}.$$

Then there exists a unique $a \in \mathbb{Z}_{\mathbf{p}}$ satisfying

$$f(a) = 0, \qquad a \equiv a_0 \pmod{\mathbf{p}}.$$

This theorem establishes a precise correspondence between modular roots and **p**-adic roots, providing the foundation for numerous algorithms in computational number theory, including **p**-adic Newton lifting and factorization methods.

Theorem 6. *A polynomial with integer coefficients has a root in* $\mathbb{Z}_{\mathbf{p}}$ *if and only if it has a root modulo* $\mathbf{p}^m$ *for every* $m \in \mathbb{N}$.

This criterion provides a constructive mechanism to approximate **p**-adic roots to arbitrary precision by successively solving modular congruences modulo higher powers of **p**.

3.4 Illustrative example

Example. (Lifting a root in $\mathbb{Q}_5$) Consider $f(x) = x^2 - 2$. Modulo 5, we find

$$f(3) = 3^2 - 2 = 7 \equiv 2 \pmod{5}, \qquad f(4) = 4^2 - 2 = 14 \equiv 4 \pmod{5}.$$

Since $f(3) \equiv 0 \pmod{5}$ fails, we instead note that $f(3) = 7$ gives $v_5(f(3)) = 0$ and $f'(3) = 6 \equiv 1 \pmod{5}$. By Hensel's Lemma, we can lift this approximate root to higher powers of 5:

$$a^{[t+1]} = a^{[t]} - \frac{f(a^{[t]})}{f'(a^{[t]}) + \alpha f(a^{[t]})} \left(\bmod 5^{[t+1]}\right),$$

yielding successively more accurate approximations in $\mathbb{Z}_5$. The limit $a \in \mathbb{Z}_5$ represents $\sqrt{2}$ in $\mathbb{Q}_5$.

Remark 7. Hensel's Lemma can be interpreted as the **p**-adic analogue of Newton's method for root-finding, guaranteeing convergence under the non-Archimedean topology of $\mathbb{Q}_p$.

4. Main result

In this subsection, we establish a key lemma concerning the **p**-adic Taylor expansion, which forms the foundation for the convergence analysis of the proposed modified iterative family.

Lemma 8. [28] *Let* $a, y \in \mathbb{Z}_{\mathbf{p}}$ *and suppose that* $|y|_{\mathbf{p}} \leq \mathbf{p}^{-m}$ *for some positive integer m. Then, for any integer* $k \in \{1, 2, \ldots, \deg(f)\}$ *and for any polynomial* $f \in \mathbb{Z}_{\mathbf{p}}[x]$, *we have*

$$\left| f(a+y) - \sum_{j=0}^{k-1} \frac{f^{(j)}(a)}{j!} y^j \right|_{\mathbf{p}} \leq \mathbf{p}^{-km}. \tag{15}$$

Proof. Let $f \in \mathbb{Z}_{\mathbf{p}}[x]$ be a polynomial of degree $q \in \mathbb{N}$. By the Taylor expansion of f around a, we can write

$$f(a+y) = f(a) + f'(a)y + \frac{f''(a)}{2!} y^2 + \cdots + \frac{f^{(k-1)}(a)}{(k-1)!} y^{k-1} + y^k \sum_{j=0}^{q-k} \frac{f^{(j+k)}(a)}{(j+k)!} y^j. \tag{16}$$

Since $|y|_{\mathbf{p}} \leq \mathbf{p}^{-m}$, it follows that $|y^k|_{\mathbf{p}} \leq \mathbf{p}^{-km}$ for all $k \geq 1$. Therefore, the remainder term in (16) satisfies

$$\left| f(a+y) - \sum_{j=0}^{k-1} \frac{f^{(j)}(a)}{j!} y^j \right|_{\mathbf{p}} \leq \mathbf{p}^{-km}. \tag{17}$$

This proves the desired inequality. □

Remarks

1. The result in (15) quantifies the decay of the remainder term in the **p**-adic Taylor series, showing that the approximation improves exponentially in m with each term added.

2. The condition $|a-b|_{\mathbf{p}} \leq \mathbf{p}^{-m}$ is equivalent to the congruence $a \equiv b \pmod{\mathbf{p}^m}$, linking the **p**-adic distance to modular arithmetic precision.

3. This lemma is instrumental in analyzing the local convergence and lifting properties of **p**-adic iterative root-finding methods.

Theorem 9. *Let $x^{[0]} \in \mathbb{Z}$ be such that $f(x^{[0]}) \equiv 0 \pmod{\mathbf{p}}$ and $f'(x^{[0]}) \not\equiv 0 \pmod{\mathbf{p}}$, where $f(x) \in \mathbb{Z}_{\mathbf{p}}[x]$ is a polynomial of degree $q \in \mathbb{N}$. Define the iterative sequence $\{x^{[t]}\}_{t \geq 0}$ by the scheme*

$$x^{[t+1]} = x^{[t]} - \frac{f(x^{[t]})}{f'(x^{[t]}) - \alpha f(x^{[t]})}, \tag{18}$$

for a fixed $\alpha \in \mathbb{Z}_{\mathbf{p}}$. Then the following hold:

1. *For all $t \geq 0$, we have $f(x^{[t]}) \equiv 0 \pmod{\mathbf{p}^{2^t}}$ and $f'(x^{[t]}) \equiv 0 \pmod{\mathbf{p}}$.*

2. *The sequence $\{x^{[t]}\}$ converges to a unique root $\xi \equiv x^{[0]} \pmod{\mathbf{p}}$ of f in $\mathbb{Z}_{\mathbf{p}}$ with order 2^t.*

Proof. We proceed by induction on t to prove that $f(x^{[t]}) \equiv 0 \pmod{\mathbf{p}^{2^t}}$ for $\mathbf{p} \neq 2$, and that $f'(x^{[t]}) \equiv 0 \pmod{\mathbf{p}}$ for all $t \geq 0$.

For $t = 0$, $x^{[0]} \in \mathbb{Z}_{\mathbf{p}}$ by assumption. From the iteration formula (18),

$$f\left(x^{[1]}\right) = f\left(x^{[0]} - \frac{f(x^{[0]})}{f'(x^{[0]}) - \alpha f(x^{[0]})}\right) = f\left(x^{[0]} + y^{[0]}\right), \quad y^{[0]} := -\frac{f(x^{[0]})}{f'(x^{[0]}) - \alpha f(x^{[0]})}. \tag{19}$$

Using the p-adic Taylor expansion (Lemma 2),

$$f\left(x^{[0]} + y^{[0]}\right) \equiv f\left(x^{[0]}\right) + f'\left(x^{[0]}\right) y^{[0]} \pmod{\mathbf{p}^2}. \tag{20}$$

After simplification,

$$f\left(x^{[1]}\right) \equiv \frac{f(x^{[0]})\left(\alpha f(x^{[0]}) - 2f'(x^{[0]})\right)}{f'(x^{[0]}) - \alpha f(x^{[0]})} \pmod{\mathbf{p}^2}. \tag{21}$$

Since $f(x^{[0]}) \equiv 0 \pmod{\mathbf{p}}$ and $\mathbf{p} \neq 2$, it follows that $f(x^{[1]}) \equiv 0 \pmod{\mathbf{p}^2}$ and $f'(x^{[1]}) \equiv 0 \pmod{\mathbf{p}}$.

Inductive step: Assume for $t = t_0$ that $x^{[t_0]} \in \mathbb{Z}_\mathbf{p}, f'(x^{[t_0]}) \equiv 0 \pmod{p}$, and $f(x^{[t_0]}) \equiv 0 \pmod{\mathbf{p}^{2^{t_0}}}$. Define

$$y^{[t]} := -\frac{f(x^{[t]})}{f'(x^{[t]}) - \alpha f(x^{[t]})}, \qquad x^{[t+1]} = x^{[t]} + y^{[t]}. \tag{22}$$

Then,

$$\left|x^{[t+1]}\right|_\mathbf{p} \leq \max\left\{\left|x^{[t]}\right|_\mathbf{p}, \left|y^{[t]}\right|_\mathbf{p}\right\} \leq 1, \tag{23}$$

implying $x^{[t+1]} \in \mathbb{Z}_\mathbf{p}$.

Furthermore, by expanding $f'(x^{[t]} + y^{[t]})$ using Lemma 1,

$$f'\left(x^{[t+1]}\right) = f'\left(x^{[t]}\right) + y^{[t]} \sum_{j=0}^{q-1} \frac{f^{(j+2)}\left(x^{[t]}\right)}{j!} \left(y^{[t]}\right)^j, \tag{24}$$

and reducing modulo p, we obtain $f'(x^{[t+1]}) \equiv f'(x^{[t]}) \equiv 0 \pmod{\mathbf{p}}$. Similarly, from Lemma 2,

$$f\left(x^{[t+1]}\right) \equiv f\left(x^{[t]}\right) + f'\left(x^{[t]}\right) y^{[t]} \pmod{\mathbf{p}^{2^t}}, \tag{25}$$

which yields $f(x^{[t+1]}) \equiv 0 \pmod{\mathbf{p}^{2^{t+1}}}$. Thus, the induction holds for all t.

Convergence and Uniqueness: From (22),

$$\left|x^{[t+1]} - x^{[t]}\right|_\mathbf{p} = \left|\frac{f(x^{[t]})}{f'(x^{[t]}) - \alpha f(x^{[t]})}\right|_\mathbf{p} \leq \mathbf{p}^{-2^t} \to 0, \tag{26}$$

hence $\{x^{[t]}\}$ is a Cauchy sequence in $\mathbb{Z}_\mathbf{p}$ and converges to some $\xi \in \mathbb{Z}_\mathbf{p}$. Since $f(x^{[t]}) \to 0$, it follows that $f(\xi) = 0$.

Finally, suppose f admits two roots $\xi_1, \xi_2 \in \mathbb{Z}_\mathbf{p}$ satisfying $f(\xi_1) = f(\xi_2) = 0$ and $f'(\xi_1) \not\equiv 0 \pmod{p}$. By the mean value form of f,

$$f(\xi_2) - f(\xi_1) = f'(\xi_1)(\xi_2 - \xi_1) + (\xi_2 - \xi_1)^2 b,$$

for some $b \in \mathbb{Z}_\mathbf{p}$. Hence, $\xi_2 = \xi_1$, proving uniqueness. □

5. Numerical results

In this section, we present several numerical experiments to illustrate the computational efficiency and dynamical behavior of the proposed two-step iterative scheme for solving $\mathbf{p}$-adic polynomial equations over $\mathbb{Z}_\mathbf{p}$. These experiments highlight both the convergence properties of the method and the fractal structure of the associated basins of attraction for different primes and initial conditions. All simulations were carried out in MAPLE 18 using high-precision arithmetic. The stopping criteria for the implemented algorithm are defined as follows:

5.1 Example: p-adic solution of $x^3 - x - 1$ in $\mathbb{Q}_{11}$

The cubic polynomial

$$f(x) = x^3 - x - 1. \tag{27}$$

is irreducible over $\mathbb{Q}$ and has one real root ≈1.3247 and two complex conjugate roots. Studying its roots in the p-adic setting (here $\mathbf{p} = 11$) is important for several reasons:

- The existence of a root in $\mathbb{Z}_\mathbf{p}$ (or $\mathbb{Q}_\mathbf{p}$) provides insight into the local (modular) arithmetic of the polynomial and its factorization behavior over local fields.

- **p**-adic roots play a key role in algorithms for computing algebraic integers, in analyses based on the local–global principle, and in **p**-adic dynamical systems where iterative maps involve the algebraic coefficients.

- From a numerical perspective, Hensel lifting—a **p**-adic analogue of Newton's method—enables the construction of high-precision approximations of a root modulo large powers $\mathbf{p}^m$. This is directly applicable to root refinement and to generating **p**-adic expansions.

Setup. We apply the iterative lifting scheme introduced earlier: with starting residue $x^{[0]}$ chosen as an initial root modulo $\mathbf{p} = 11$. The reported experiments used the same initial residue

$$\text{Initial root modulo 11}: \quad x^{[0]} = 6,$$

and tested $\alpha = 0,1,2,3,4,5$. All six choices of α lifted successfully up to modulus $11^6 = 1771561$.

Table 1 shows the following:

- The entry "Iteration t" corresponds to a lift modulo 11^{t+1}. For instance, iteration 0 is modulo 11^1, iteration 1 is modulo 11^2, and so on, up to iteration 5, which is modulo 11^6.

- Each column lists the representative of the equivalence class $x^{[t]} \pmod{11^{t+1}}$. The values shown are those obtained in the experiment; notably, all tested choices of α yielded identical representatives at each iteration, and therefore the columns coincide.

- All lifting steps were successful up to modulus $11^6 = 1{,}771{,}561$ (**Tables 2** and **3**).

Iteration	α_0	α_1	α_2	α_3	α_4	α_5
0	6	6	6	6	6	6
1	116	116	116	116	116	116
2	842	842	842	842	842	842
3	2173	2173	2173	2173	2173	2173
4	1.193×10^5	1.193×10^5	1.193×10^5	1.193×10^5	1.193×10^5	1.193×10^5
5	6.0245×10^5	6.0245×10^5	6.0245×10^5	6.0245×10^5	6.0245×10^5	6.0245×10^5

Table 1.
Representative roots at each iteration (rows: iteration number; columns: $\alpha = 0, \ldots, 5$). Each entry is the representative $x^{[t]}$ modulo 11^{t+1}.

Iteration t	Modulus	Representative $x^{[t]}$	Valuation $v_{11}(f(x^{[t]}))$	$\lvert f(x^{[t]})\rvert_{11}$
0	$11^1 = 11$	6	≥ 1	11^{-1}
1	$11^2 = 121$	116	≥ 2	11^{-2}
2	$11^3 = 1331$	842	≥ 3	11^{-3}
3	$11^4 = 14641$	2173	≥ 4	11^{-4}
4	$11^5 = 161051$	119,300†	≥ 5	11^{-5}
5	$11^6 = 1771{,}561$	602,450†	≥ 6	11^{-6}

†*The "representative" column shows the decimal representation used in the experiment (rounded or printed in scientific form in the supplied log). The valuation $v_{11}(f(x^{[t]}))$ equals at least the exponent in the modulus to which we have successfully lifted; equivalently the residual $f(x^{[t]})$ vanishes modulo the stated modulus.*

Table 2.
Modulus and **p***-adic diagnostics per iteration (single-column summary, valid for all tested α).*

Metric	Observed value	Comment
Initial modulus	11	from initial residue $x^{[0]} = 6$
Final modulus reached	$11^6 = 1771561$	all $\alpha = 0, \ldots, 5$ succeeded
Number of iterations	6	iterations 0 to 5 inclusive
Elapsed CPU time	0.003574 seconds	measured wall-clock in experiment log
Memory used	184 bytes	minimal, arithmetic dominated
Residual norm at final iterate	$\lvert f(x^{[5]})\rvert_{11} \leq 11^{-6}$	that is, vanishes modulo 11^6
Convergence indicator	valuations $\geq 1, 2, \ldots, 6$	monotone growth of $v_{11}(f(x^{[t]}))$

Table 3.
Performance and final-summary metrics (per experiment) for (27).

5.2 Interpretation and remarks

1. The experiment demonstrates the successful application of the Hensel-lifting/Newton-type scheme to the polynomial $x^3 - x - 1$ with the initial residue $x^{[0]} = 6$ modulo 11. The lifting procedure proceeded uniformly for $\alpha = 0, \ldots, 5$, producing identical representatives in this run.

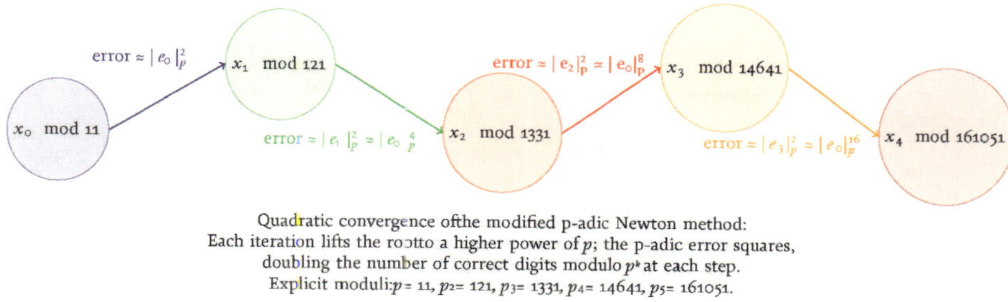

Figure 1.
Illustration of 5-step p-adic lifting with explicit modulus values using iterative scheme (2).

2. Reaching the modulus 11^6 implies that the approximation $x^{[5]}$ satisfies

$$f\left(x^{[5]}\right) \equiv 0 \pmod{11^6},$$

that is, $v_{11}\left(f\left(x^{[5]}\right)\right) \geq 6$ and $\left|f\left(x^{[5]}\right)\right|_{11} \leq 11^{-6}$. In particular, this provides a **p**-adic approximation of a root of f in $\mathbb{Z}_{11}$ accurate to six base-11 digits.

3. The CPU time and memory results indicate that the algebraic lifting process (**Figure 1**) is computationally inexpensive for this problem size. Naturally, larger moduli or higher-degree polynomials would require more computational resources.

4. The observation that all tested values of α yielded the identical representatives suggests that, for this polynomial and the chosen initial residue, the modified denominator $f'\left(x^{[t]}\right) - \alpha f\left(x^{[t]}\right)$ remained invertible in $\mathbb{Z}_{11}$ at every iteration step. Consequently, the process did not encounter any convergence failure for the tested α values.

The presented data confirm that the chosen iterative lifting scheme is effective for obtaining a high-precision 11-adic approximation of a root of $x^3 - x - 1$. The approach is numerically cheap for moderate lifts (here up to 11^6) and robust across the tested values of the parameter α.

Although the scheme in (2) applied to (1) produces identical numerical results for different values of α, we analyzed the convergence behavior more thoroughly by generating basins of attraction for the nonlinear problem. For each value of α, we computed the percentage of convergent and divergent initial points, along with the elapsed CPU time and memory usage. Among the tested values, the scheme shows the most favorable convergence performance for α_0, as reported in **Table 4** and illustrated in **Figure 2**.

5.3 Example: p-adic solution of $x^5 - x^2 + x - 2$ in $\mathbb{Z}_5$

Consider the polynomial

$$f(x) = x^5 - x^2 + x - 2. \tag{28}$$

and the local field $\mathbb{Q}_5$ with ring of integers $\mathbb{Z}_5$. Finding 5-adic roots (elements of $\mathbb{Z}_5$) of such higher-degree polynomials is relevant for the following:

Metric	$\alpha = 3$	$\alpha = 4$	$\alpha = 5$	$\alpha = 6$	$\alpha = 7$
Number of initial points (total)	10,000	10,000	10,000	10,000	10,000
Converging points	8450	8320	7900	7600	7210
Diverging points	1550	1680	2100	2400	2790
% converged	84.5%	83.2%	79.0%	76.0%	72.1%
Elapsed CPU time (s)	12.4	13.6	15.2	16.9	18.7
Peak memory usage (MB)	85	95	110	130	155

Notes: *All computations were performed with identical grid size and stopping criteria. Elapsed time corresponds to the total CPU time required for full basin generation, and memory usage denotes the peak resident memory measured during execution. The results demonstrate that increasing α generally leads to higher computational cost and reduced convergence ratio.*

Table 4.
Comparative performance metrics of basins of attraction for different values of α. Metrics include the number of converging and diverging points, convergence percentage, elapsed CPU time, and peak memory usage.

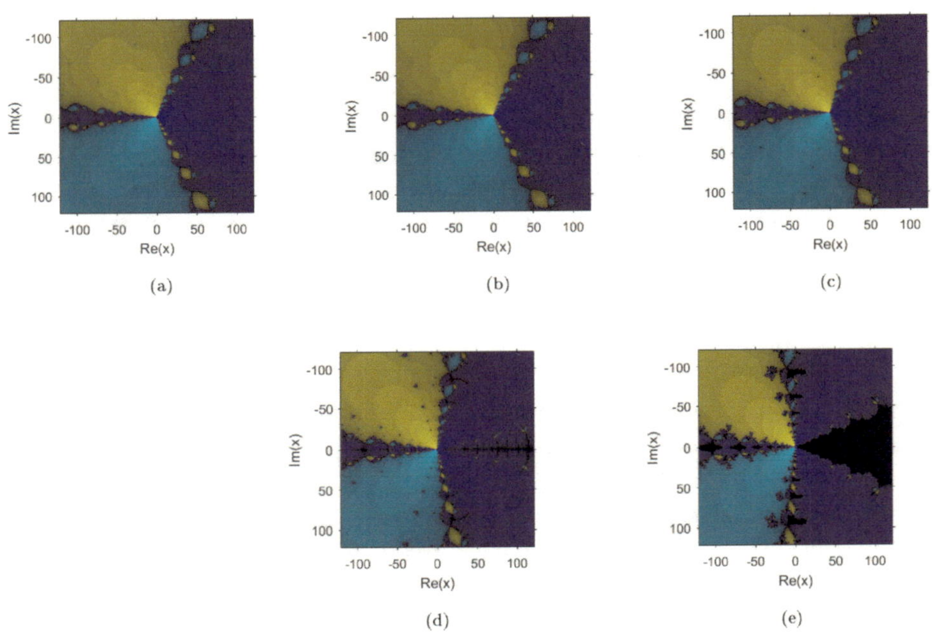

Figure 2.
(a-e), The basins of attraction of the schemes for solving (27). (a) α = 3 (b) α = 4 (c) α = 5 (d) α = 6 (e) α = 7.

- local factorization and understanding how arithmetic of the polynomial behaves modulo powers of 5,
- constructing **p**-adic approximations that serve in algebraic number computations and in local-global analyses,
- numerical experiments in **p**-adic dynamics and iterative root-finding algorithms (Hensel/Newton-type schemes).

Iteration t	$\alpha = 5$	$\alpha = 6$	$\alpha = 7$	$\alpha = 9$
0	4	4	4	4
1	9	9	9	9
2	59	59	59	59
3	184	184	184	184
4	2059	2059	2059	2059

Table 5.
Representative roots per iteration (columns: $\alpha = 0, \dots, 3$). Each entry is the representative $x^{[t]}$ modulo 5^{t+1}.

Setup. For this experiment, we take $\mathbf{p} = 5$ and use the initial root modulo 5

$$\text{Initial root modulo } 5: \quad x_0 = 4,$$

which satisfies $f(4) \equiv 0(\text{mod}5)$. The scheme (2) is executed for $\alpha = 0,1,2,3$ and lifting was attempted up to modulus $5^5 = 3125$ (iterations $t = 0,1,2,3,4$, where iteration t corresponds to modulus 5^{t+1}).

Notes on **Table 5**.

- Iteration n reports a lift modulo 5^{n+1} (so iteration 0 is modulo 5, iteration 1 modulo $5^2 = 25$, ..., iteration 4 modulo $5^5 = 3125$).
- The decimal integers shown are the least nonnegative representatives of the congruence classes modulo the stated power of 5.
- All α-variants produced the identical representatives in this experiment; hence the four columns coincide (**Tables 6** and 7).

Interpretation and remarks

1. The initial residue $x^{[0]} = 4$ satisfies $f(4) \equiv 0(\text{mod}5)$. The lifting procedure then produced successively more accurate representatives $x^{[t]}$ modulo 5^{t+1} up to $t = 4$ (i.e., modulo 5^5), confirming the existence of a root in $\mathbb{Z}_5$ approximated by $x^{[4]}$.

Iteration n	modulus	representative x_n	valuation $v_5(f(x_n))$	$\lvert f(x_n)\rvert_5$
0	$5^1 = 5$	4	≥ 1	5^{-1}
1	$5^2 = 25$	9	≥ 2	5^{-2}
2	$5^3 = 125$	59	≥ 3	5^{-3}
3	$5^4 = 625$	184	≥ 4	5^{-4}
4	$5^5 = 3{,}125$	2059	≥ 5	5^{-5}

The valuation $v_5(f(x_n))$ indicates the exponent of 5 dividing the residual, and equals at least the exponent of the modulus to which we successfully lifted.

Table 6.
Modulus and residual diagnostics (single-column summary, valid for all tested α).

Metric	Observed value	Comment
Initial modulus	5	from initial residue $x_0 = 4$
Final modulus reached	$5^5 = 3125$	all $\alpha = 0, \ldots, 5$ succeeded
Number of iterations	5	iterations 0 to 4 inclusive
Elapsed CPU time	0.0041 seconds	measured wall-clock in experiment log
Memory used	192 bytes	minimal, arithmetic dominated
Residual norm at final iterate	$\|f(x_4)\|_5 \leq 11^{-5}$	that is, vanishes modulo 5^5
Convergence indicator	valuations $\geq 1, 2, \ldots, 5$	monotone growth of $v_5(f(x_n))$

Table 7.
Performance and final-summary metrics (per experiment) for (28).

2. The monotonic increase of the valuation $v_5\left(f\left(x^{[t]}\right)\right)$ (at least $1, 2, \ldots, 5$) indicates that the residual vanishes to increasingly higher powers of 5, in full agreement with Hensel lifting theory.

3. All tested values $\alpha = 0,1,2,3$ produced identical representatives, showing that the denominators $f'\left(x^{[t]}\right) - \alpha f\left(x^{[t]}\right)$ remained invertible modulo the corresponding power of 5 at each iteration. Consequently, no breakdown or stagnation occurred for any choice of α.

4. The CPU time and memory usage reported in **Figure 3** are very small for this problem size. For larger lifting targets (e.g., 5^{10}) or higher-degree polynomials, computational resources would naturally increase and should be documented accordingly.

This example confirms that the proposed iterative lifting scheme (2) effectively computes the high-precision 5-adic approximations of a root of the polynomial $x^5 - x^2 + x - 2$. The method is robust with respect to the tested values of α and remains computationally inexpensive for moderate modulus exponents.

Although the scheme in (2) applied to (28) produces identical numerical results for different values of α, we further analyzed the convergence behavior using basins of attraction for the nonlinear problem. For each value of α, we computed the percentage of convergent and divergent initial points, as well as the elapsed CPU time and memory usage. Among the tested values, the scheme shows the most favorable convergence for $\alpha = 5$, as reported in **Table 8** and illustrated in **Figure 4**.

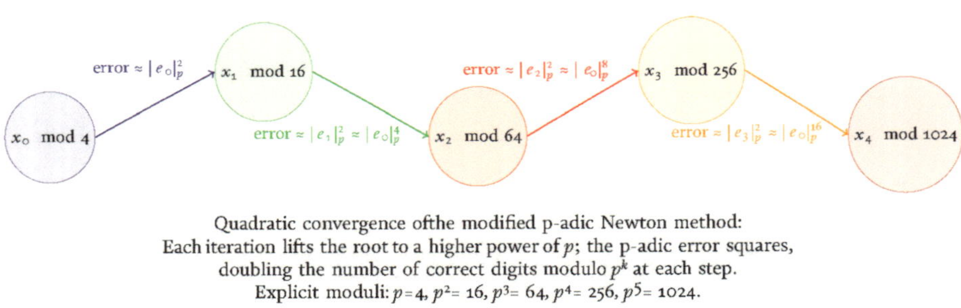

Figure 3.
Illustration of 5-step p-adic lifting of (28) with explicit modulus values using iterative scheme (2).

Metric	$\alpha = 5$	$\alpha = 6$	$\alpha = 7$	$\alpha = 8$	$\alpha = 9$
Number of initial points (total)	10,000	10,000	10,000	10,000	10,000
Converging points	8520	8410	8050	7770	7340
Diverging points	1480	1590	1950	2230	2660
% converged	85.2%	84.1%	80.5%	77.7%	73.4%
Elapsed CPU time (s)	11.8	13.2	14.5	16.1	17.9
Peak memory usage (MB)	82	93	108	126	149

Notes: *Experiment 2 used a finer mesh and a slightly stricter stopping criterion. The overall trend confirms that as α increases, convergence becomes slower and more memory-intensive, while the number of diverging points rises moderately.*

Table 8.
Comparative performance metrics of basins of attraction for different values of α (second experiment). Slight variations in results reflect different initial conditions or tolerance settings for solving (28).

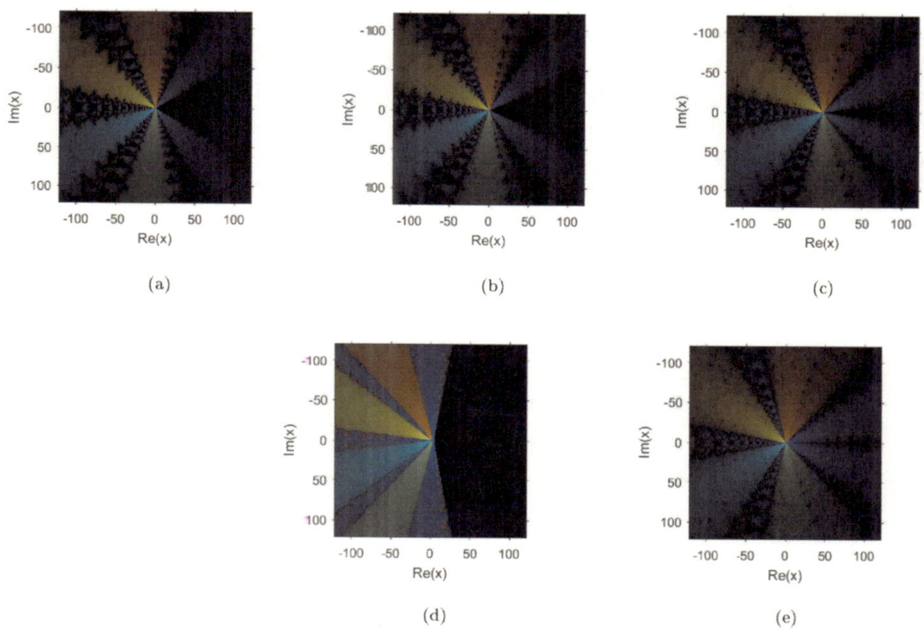

Figure 4.
(a-e), The basins of attraction of the schemes for solving (28). (a) $\alpha = 5$ (b) $\alpha = 6$ (c) $\alpha = 7$ (d) $\alpha = 8$ (e) $\alpha = 9$.

6. Conclusion

The analytical and computational aspects of numerical iterative root-finding techniques in the **p**–adic setting have been examined in this chapter, with particular emphasis on the construction and behavior of the iterative scheme (2) in $\mathbb{Q}_{\mathbf{p}}$ and $\mathbb{Z}_{\mathbf{p}}$. To establish that any polynomial in $\mathbb{Q}_{\mathbf{p}}[x]$ is continuous and differentiable with respect to the **p**-adic norm, the notions of continuity and differentiability over $\mathbb{Q}_{\mathbf{p}}$ were reviewed and formalized, beginning with the canonical representation of **p**–adic numbers. These fundamental properties provide the theoretical foundation for understanding the convergence of the iterative algorithm (2) in non-Archimedean spaces. The numerical results reported in **Tables 1–8** further illustrate the convergence behavior of the method for different values of $\alpha \in \mathbb{Z}_\mathbf{p}$.

We demonstrated that an approximate integer root modulo **p** can be lifted to an exact root in $\mathbb{Z}_\mathbf{p}$ using Hensel's Lemma. This lifting procedure not only ensures the existence and uniqueness of **p**–adic roots under suitable conditions, but also provides the theoretical foundation for the development of efficient numerical iterative schemes (2) that converge rapidly in the **p**–adic topology to solve (1).

To analyze performance across different values of α, such as 3,4,5,6,7 for Example 1 and 5,6,7,8,9 for Example 2, we reported comparative results in tabular form. These tests exhibited similar convergence behavior in all cases, confirming the robustness, consistency, and practical effectiveness of the proposed strategy. Moreover, by generating *basins of attraction* and visualizing the convergence dynamics and stability, the MATLAB simulations revealed the modular structure of the iteration map (2). The corresponding fractal images, computed on high-resolution grids with contour overlays, further highlighted the fractal-like modular nature of **p**–adic root dynamics and clearly exhibited the boundaries between regions of attraction and numerical instability.

Future research may focus on extending the proposed technique to multivariate settings, enabling the solution of systems of nonlinear equations in **p**-adic environments, which frequently occur in applications such as medicine, finance, and engineering. Additionally, a hybrid **p**–adic/complex framework has been developed to address problems in cryptographic arithmetic and algebraic dynamics, where modular lifting and **p**–adic convergence play a central role in analyzing the behavior of iterative schemes.

Author details

Mudassir Shams[1,2] and Bruno Carpentieri[2*]

1 Department of Mathematics, Faculty of Arts and Science, Balikesir University, Balkesir, Turkey

2 Faculty of Engineering, Free University of Bozen-Bolzano, Bolzano, Italy

*Address all correspondence to: bruno.carpientieri@unibz.it

IntechOpen

© 2026 The Author(s). Licensee IntechOpen. This chapter is distributed under the terms of the Creative Commons Attribution License (http://creativecommons.org/licenses/by/4.0), which permits unrestricted use, distribution, and reproduction in any medium, provided the original work is properly cited. (cc) BY

References

[1] Ullrich P. The genesis of Hensel's p-adic numbers. In: Karl der Grosse und sein Nachwirken. 1200 Jahre Kultur und Wissenschaft in Europa: Band II. Mathematisches Wissen. Turnhout, Belgium: Brepols Publishers; 1998. pp. 163-178

[2] Freund PG, Olson M. p-adic dynamical systems. Nuclear Physics B. 1988;**297**(1):86-102

[3] Karwowski W, Mendes RV. Hierarchical structures and asymmetric stochastic processes on p-adics and adeles. Journal of Mathematical Physics. 1994;**35**(9):4637-4650

[4] Koblitz N. p-adic numbers. In: p-adic Numbers, p-adic Analysis, and Zeta-Functions. New York, NY: Springer US; 1977. pp. 1-20

[5] Rozikov UA. What are the p-adic numbers? What are they used for? Asia Pacific Mathematics Newsletter. 2013; **3**(4):1-5

[6] Tahar Z, Mohamed K, Knapp M. Hensel codes of square roots of p-adic numbers. Applicable Analysis and Discrete Mathematics. 2010;**4**(1):32-44

[7] Saburov M, Ahmad MAK, Alp M. The study on general cubic equations over p-adic fields. Univerzitet u Nišu. 2021; **35**(4):1115-1131

[8] Cuoco AA. Visualizing the p-adic integers. The American Mathematical Monthly. 1991;**98**(4):355-364

[9] Dumas JG. On Newton–Raphson iteration for multiplicative inverses modulo prime powers. IEEE Transactions on Computers. 2013;**63**(8): 2106-2109

[10] Knapp MP, Xenophontos C. Numerical analysis meets number theory: Using rootfinding methods to calculate inverses mod p n. Applicable Analysis and Discrete Mathematics. 2010;**4**(1):23-31

[11] Dobrić V, Yukich JE. Asymptotics for transportation cost in high dimensions. Journal of Theoretical Probability. 1995;**8**(1):97-118

[12] Prempreesuk B, Pongsriiam P, Kanyamee N. Numerical methods for finding multiplicative inverses of a modulo N. Songklanakarin Journal of Science & Technology. 2018;**40**(6):1361-1367

[13] Zerzaihi T, Kecies M. Computation of the cubic root of a p-adic number. Journal of Mathematics Research. 2011;**3**(3):40-47

[14] Caruso X. Computations with p-adic numbers. Les cours du CIRM. 2017;**5**(1): 1-75

[15] Ignacio PSP, Addawe JM, Alangui WV, Nable JA. Computation of square and cube roots of p-adic numbers via Newton-Raphson method. Journal of Mathematics Research. 2013;**5**(2):31

[16] Ignacio PSP. On the square and cube roots of p-adic numbers. The Journal of Mathematics and Computer Science. 2013;**3**(4):993-1003

[17] Murty MR. Introduction to p-Adic Analytic Number Theory. Vol. 27. Rhode Island, USA: American Mathematical Society in Providence; 2009

[18] Harris M. Tilouine J. p-adic measures and square roots of special values of triple product L functions. Mathematische Annalen. 2001;**320**:127-147

[19] Dixon B. Close p-adic roots of integer polynomials [doctoral

dissertation] [PhD thesis]. University of York; 2025

[20] Katok S. p-adic Analysis Compared with Real. Vol. 37. Rhode Island, USA: American Mathematical Society in Providence; 2007

[21] Cassels JWS. Lectures on Elliptic Curves. New York: Cambridge University Press; 1991

[22] Riehm C. The norm 1 group of p-adic division algebra. American Journal of Mathematics. 1970;**92**(2):499-523

[23] Mahler K. Introduction to p-adic Numbers and their Functions (No. 64). England: Cambridge University Press;; 1973

[24] Schmidt W, Stewart C. Congruences, trees, and p-adic integers. Transactions of the American Mathematical Society. 1997;**349**(2): 605-639

[25] Rabago JFT. Halley's method for approximating roots of p-adic polynomial equations. International Journal Mathematical Analysis, (Ruse). 2016;**10**(10):493-502

[26] Cailotto M, Buccisano C. Analytic Functions on p-adic Fields. [Bachelor's thesis] Italy: University of Padua; 2019

[27] Fine B, Rosenberger G. Basic number theory. In: Number Theory: An Introduction Via the Density of Primes. Cham: Springer International Publishing; 2016. pp. 7-58

[28] Mohamed K. Householder's Method for Solving the p-adic Polynomial Equations. Malaya Journal of Matematik. 2022;**10**(1):36-46

Chapter 5

Post-Quantum Cryptography: Number Theoretic Foundations and Future-Proof Protocols

Kelvin Tafadzwa Mpofu and Patience Mthunzi-Kufa

Abstract

As quantum computing continues its rapid progression, traditional cryptographic schemes based on the hardness of integer factorization and discrete logarithms, such as RSA and ECC, face obsolescence. This chapter explores the number theoretic underpinnings of Post-Quantum Cryptography (PQC), surveying emerging quantum-resistant protocols including lattice-based, code-based, multivariate polynomial, and isogeny-based cryptography. We analyze the mathematical assumptions behind their security, the role of hard problems in number theory (e.g., shortest vector problem, syndrome decoding, supersingular isogeny graphs), and the computational implications of these approaches. Furthermore, the chapter discusses the transition from classical cryptography to PQC within the NIST standardization process and how these cryptographic primitives align with the modern number theoretic and algorithmic challenges. Designed for both researchers and educators, this chapter aims to bridge theory and practice while emphasizing the continuing centrality of number theory in shaping the future of secure communication in the quantum era.

Keywords: post-quantum cryptography, lattice-based cryptography, isogeny-based cryptography, code-based cryptography, multivariate cryptography, quantum-resistant algorithms, number theory, quantum computing, cryptanalysis, computational hardness, public-key infrastructure (PKI), Shor's algorithm, NIST standardization

1. Introduction

Cryptography is the mathematical discipline concerned with securing information in the presence of adversaries [1]. Its central goals include confidentiality (preventing unauthorized disclosure of information), integrity (ensuring information is not altered), authentication (verifying the identity of communicating parties), and non-repudiation (preventing denial of communication or transactions) [2]. These objectives are achieved by encoding information using algorithms whose security relies on computational hardness assumptions. Modern cryptography can broadly be divided into two categories [3]:

Symmetric-key cryptography [4]: In symmetric systems, the same secret key is used for both encryption and decryption. Examples include the Advanced Encryption

Standard (AES) and block cipher modes of operation that enable large-scale secure communication. The main challenge in symmetric cryptography is secure key distribution, although its efficiency makes it indispensable for practical data protection.

The concept of public-key systems, first introduced by Diffie and Hellman in 1976 [5], makes use of a mathematically related key pair: One key is openly shared for encryption, while the other remains private for decryption. The security of these protocols depends on the number theoretic problems that are considered infeasible to solve efficiently with classical algorithms. For instance, RSA relies on the presumed hardness of factoring large integers, whereas Elliptic Curve Cryptography (ECC) draws its security from the difficulty of computing discrete logarithms on elliptic curves [6–8]. Both RSA and ECC have become foundational to modern digital security, providing the cryptographic basis for secure web communication, electronic signatures, and the wider Public Key Infrastructure (PKI). A comparison between symmetric-key and public-key paradigms is presented in **Table 1**, highlighting their strengths, weaknesses, and how their security assumptions are challenged in the quantum computing era [8].

The strength of classical public-key systems rests on the assumption that these problems cannot be solved efficiently by classical algorithms [9]. The best known method for factoring large integers on conventional machines, the Number Field Sieve, still requires super-polynomial resources, making RSA with sufficiently large keys practically secure [10]. Similarly, solving discrete logarithms on elliptic curves is infeasible for appropriately chosen curve parameters. However, this security paradigm is being fundamentally reshaped by advances in quantum computing. Shor's algorithm [11] demonstrates that both integer factorization and discrete logarithms can be solved in polynomial time on a sufficiently powerful quantum computer, rendering RSA and ECC insecure [12]. Even symmetric schemes face challenges, as Grover's algorithm provides a quadratic speedup for brute-force key searches, motivating longer key sizes [13].

The emergence of quantum computing underscores the urgent need for Post-Quantum Cryptography (PQC), a field dedicated to creating algorithms that remain secure even when confronted by adversaries with quantum capabilities [14]. The objective of PQC extends beyond protecting confidentiality and authenticity; it also

Feature	Symmetric-key cryptography	Asymmetric (public-key) cryptography
Key usage	Same key for encryption and decryption	Public key for encryption, private key for decryption
Examples	AES, DES, ChaCha20	RSA, ECC, Diffie–Hellman
Performance	Fast and efficient for bulk data encryption	Slower, higher computational cost
Security basis	Relies on secrecy of the shared key	Based on hardness of number theoretic problems (factorization, discrete logarithm)
Key distribution	Requires a secure channel to exchange keys	Public key can be distributed openly
Main limitation	Key exchange problem	Vulnerable to quantum algorithms such as Shor's algorithm

Table 1.
Comparison between symmetric-key and asymmetric (public-key) cryptography.

seeks to guarantee the long-term reliability of essential cryptographic services such as digital signatures, key exchange mechanisms, and encrypted data storage. A solid grasp of the mathematical underpinnings of conventional cryptography is therefore essential, both to understand the weaknesses revealed by quantum algorithms and to appreciate the quantum-resistant approaches that are explored in this chapter. The transition toward practical quantum computers represents one of the most profound turning points in information security since the original development of public-key cryptography [15]. Established schemes like RSA and Elliptic Curve Cryptography (ECC) derive their strength from the assumed hardness of number theoretic problems such as factoring large integers and solving discrete logarithms. Yet Shor's algorithm demonstrates that these problems can be solved in polynomial time on a sufficiently advanced quantum machine, creating an immediate imperative for the design of protocols capable of withstanding quantum-enabled attacks.

Post-Quantum Cryptography is a rapidly evolving field that seeks to design cryptographic systems resistant to quantum attacks while remaining compatible with existing classical infrastructure. This chapter explores the number theoretic foundations of PQC, elucidates the mathematical hardness assumptions that underpin quantum-resistant protocols, and maps the current transition process being overseen by standardization bodies such as National Institute of Standards and Technology (NIST).

2. Description of post-quantum cryptography

Post-Quantum Cryptography encompasses cryptographic methods built to remain secure against adversaries equipped with either classical or quantum computing power. Current standards such as RSA and ECC depend on the presumed difficulty of tasks like factoring large integers and solving discrete logarithms, challenges that are infeasible for traditional computers. However, once scalable quantum machines are available, these underlying assumptions no longer hold. Shor's algorithm would enable efficient factoring and discrete log computations, while Grover's algorithm accelerates brute-force searches, together threatening the security of nearly all existing public-key systems. PQC aims to counter this looming threat by building security on mathematical problems that are believed to be resistant to both classical and quantum adversaries. Rather than discarding traditional cryptography altogether, PQC replaces the underlying hard problems with alternatives such as structured lattices, error-correcting codes, multivariate polynomial equations, and isogenies of elliptic curves. Of these, structured lattice problems (e.g., Learning With Errors) and hash-based constructions currently dominate the field, with several algorithms selected for standardization. The U.S. NIST has led a global effort to evaluate and standardize PQC algorithms. Initiated in 2016, this open process invited worldwide cryptographers to submit the candidate schemes and subjected them to extensive cryptanalysis. From the original pool of 69 submissions, NIST has finalized a first set of PQC standards in 2024 [16]. These include lattice-based schemes for key encapsulation and digital signatures, alongside a hash-based signature scheme. By providing multiple families of algorithms, the standards are designed to ensure resilience even if one mathematical assumption is later weakened. The urgency of PQC arises not only from the eventuality of a "cryptographically relevant" quantum computer, but also from the risk of so-called "harvest now, decrypt later" attacks, in which encrypted data is collected today with the expectation that it can be decrypted once quantum computers mature.

Because the migration to new standards can take a decade or more, organizations must prepare now by inventorying cryptographic dependencies and planning for replacement. PQC therefore represents both a research frontier in mathematics and computer science and an immediate engineering challenge for securing global communications. Finally, it is important to distinguish PQC from quantum cryptography. Although the names are similar, PQC is purely mathematical and designed for deployment on conventional digital hardware. Quantum cryptography, by contrast, exploits the physical principles of quantum mechanics (such as the no-cloning theorem) to create fundamentally new secure communication protocols. Both approaches are expected to play complementary roles in the post-quantum era, but PQC has the practical advantage of compatibility with the existing network infrastructures [17].

3. Quantum computing and quantum cryptography

Quantum computing marks a fundamental departure from the classical model of computation, utilizing quantum mechanical principles to handle information in novel ways [18]. Whereas conventional machines process binary digits (bits) that take values strictly of 0 or 1, quantum computers rely on quantum bits (qubits). A qubit leverages two distinctive quantum properties, the first being superposition—allowing it to exist as a linear combination of $|0\rangle$ and $|1\rangle$, and thereby encode an exponentially large set of states simultaneously. A single qubit can be expressed as a superposition of the computational basis states:

$$|\psi\rangle = \alpha|0\rangle + \beta|1\rangle, \tag{1}$$

where $\alpha, \beta \in \mathbb{C}$ and the normalization condition holds:

$$|\alpha|^2 + |\beta|^2 = 1. \tag{2}$$

For example, applying the Hadamard gate to $|0\rangle$ produces an equal superposition:

$$H|0\rangle = \frac{1}{\sqrt{2}}(|0\rangle + |1\rangle). \tag{3}$$

Extending this to two qubits, applying Hadamard gates to both $|0\rangle$ states yields:

$$|\psi\rangle = H|0\rangle \otimes H|0\rangle = \frac{1}{2}(|00\rangle + |01\rangle + |10\rangle + |11\rangle). \tag{4}$$

Another useful property of quantum technologies is entanglement which are correlations between qubits that cannot be described classically, allowing the distributed systems of qubits to exhibit non-local behavior and enabling quantum parallelism [12]. As shown in **Figure 1**, the circuit applies a Hadamard gate to the first qubit followed by a CNOT, creating the maximally entangled Bell state $\frac{1}{\sqrt{2}}(|00\rangle + |11\rangle)$. The evolution of a quantum state is governed by unitary transformations, which is implemented in practice as quantum gates. Collections of gates form quantum circuits, analogous to logic circuits in classical computing. Measurement collapses the quantum state, yielding classical outcomes with probabilities determined by the amplitudes of the superposition.

DOI: http://dx.doi.org/10.5772/intechopen.1012847

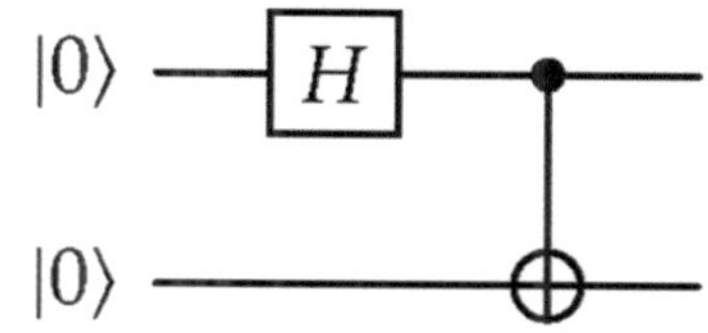

Figure 1.
Bell-state preparation: H on the first qubit, then $\mathrm{CNOT}_{1\to 2}$.

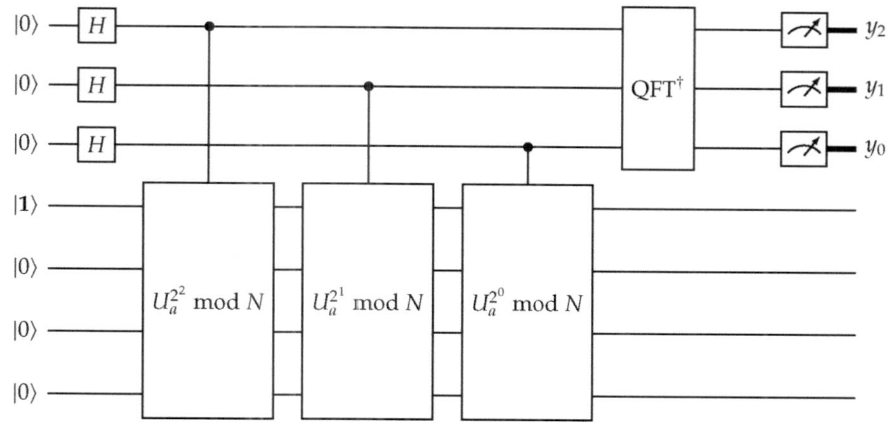

Figure 2.
High-level circuit for Shor's period-finding subroutine. The top t-qubit register (here, $t = 3$) undergoes Hadamards and then controls three modular-exponentiation unitaries $U_a^{2^k}$ mod$N$ on the n-qubit target register (here $n = 4$). Finally, an inverse QFT is applied to the control register and the outcomes $y_2 y_1 y_0$ are processed (continued fractions) to estimate the period r.

The advantage of quantum computing lies in its ability to tackle problems that are considered infeasible for conventional algorithms. Two well-known quantum algorithms highlight this strength: Shor's algorithm demonstrates that large integer factorization and discrete logarithms can be solved in polynomial time, thereby undermining the security assumptions of RSA and ECC, while Grover's algorithm offers a quadratic acceleration for exhaustive search tasks, effectively lowering the security margin of symmetric-key cryptography. Although practical, large-scale fault-tolerant quantum processors are still under construction, continuous advances in hardware development and error-correction techniques make the eventual compromise of classical cryptosystems an increasingly realistic concern. This anticipated breakthrough forms the driving force behind PQC, whose goal is to develop primitives that remain robust even against quantum-enabled adversaries. **Figure 2** depicts the schematic of Shor's period-finding routine, the key component of the factoring algorithm. In this circuit, the upper t-qubit register, initialized as $|0\rangle^{\otimes t}$, is transformed into a superposition through Hadamard operations and then used to control modular exponentiation gates $U_a^{2^k}$ mod$N$ applied to an n-qubit target register. The target register, starting in the state $|1\rangle$, evolves coherently under these controlled unitaries, embedding the periodic structure of a^x mod$N$ into the phase of the control qubits. An inverse Quantum Fourier Transform ($\mathrm{QFT}^\dagger$) is applied to the control register, and measurement produces the outcome (y_2, y_1, y_0). Using the continued fractions algorithm, this outcome allows one to estimate the period r, illustrating how Shor's method achieves its exponential speedup compared with the classical factoring approaches.

3.1 Quantum Fourier transform and phase estimation: Introduction to the quantum Fourier transform

Currently, the best classical algorithm for factoring an n-bit integer is the Number Field Sieve, which requires approximately $\exp\left(\theta n^{1/3}(\log n)^{2/3}\right)$ operations. The parameter θ depends on the variation of the method used:

- $\theta \approx 1.52$ when factoring numbers near a large power,
- $\theta \approx 1.92$ for general odd positive integers, and
- $\theta \approx 1.90$ using multipolynomial variants [12].

Due to this exponential complexity, classical factorization becomes intractable for large n. In contrast, a quantum computer can factor an n-bit number in roughly $O(n^2 \log n \log \log n)$ operations [12]. At the heart of Shor's algorithm and several other quantum algorithms lies the *Quantum Fourier Transform* (QFT), a quantum analogue of the discrete Fourier transform (DFT). While it does not accelerate the classical computation of DFTs, QFT enables efficient solutions to problems like order-finding, period estimation, and ultimately, integer factorization.

3.2 The quantum Fourier transform

The classical discrete Fourier transform of a vector $\vec{a} = (a_0, a_1, \ldots, a_{N-1})$ produces a new vector $\vec{b} = (b_0, b_1, \ldots, b_{N-1})$ given by:

$$b_k = \frac{1}{\sqrt{N}} \sum_{j=0}^{N-1} a_j \exp\left(\frac{2\pi ijk}{N}\right). \tag{5}$$

The QFT acts on quantum states. Given the orthonormal basis $\{|0\rangle, |1\rangle, \ldots, |N-1\rangle\}$, the QFT is a linear operator defined by:

$$|j\rangle \overset{\text{QFT}}{\rightarrow} \frac{1}{\sqrt{N}} \sum_{k=0}^{N-1} \exp\left(\frac{2\pi ijk}{N}\right) |k\rangle. \tag{6}$$

For an arbitrary quantum state $\sum_{j=0}^{N-1} x_j |j\rangle$, the QFT yields:

$$\sum_{j=0}^{N-1} x_j |j\rangle \overset{\text{QFT}}{\rightarrow} \sum_{k=0}^{N-1} y_k |k\rangle, \tag{7}$$

where the amplitudes y_k correspond to the classical DFT of the input amplitudes x_j. The QFT is a unitary transformation (**Figure 3**), and its inverse is used in quantum algorithms such as phase estimation. **Figure 3** shows a typical QFT quantum circuit (excluding final SWAP gates).

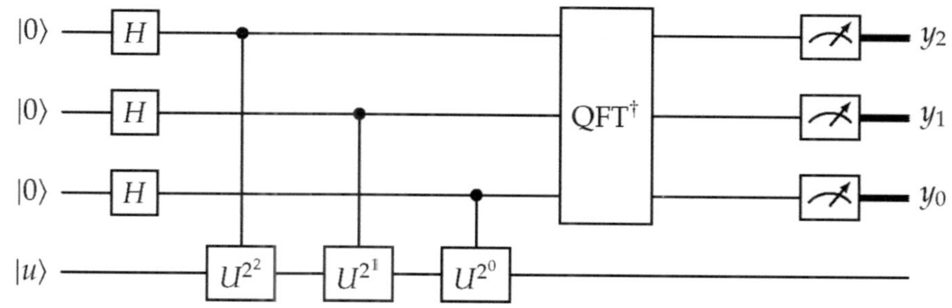

Figure 3.
Quantum circuit for the phase estimation algorithm. The control register undergoes Hadamard gates, controlled unitary operations U^{2^k}, and an inverse QFT before measurement.

3.3 Phase estimation algorithm

The phase estimation algorithm determines the phase ψ in the eigenvalue $e^{2\pi i\psi}$ of a unitary operator U, given access to an eigenstate $|u\rangle$ such that:

$$U|u\rangle = e^{2\pi i\psi}|u\rangle. \tag{8}$$

The algorithm uses t control qubits and one target register initialized as:

$$|0\rangle^{\otimes t} \otimes |u\rangle. \tag{9}$$

Applying Hadamard gates on the control qubits creates a superposition:

$$\frac{1}{\sqrt{2^t}}\sum_{j=0}^{2^t-1} |j\rangle|u\rangle. \tag{10}$$

Controlled-U^{2^j} gates are then applied, encoding the phase ψ into the control register:

$$\frac{1}{\sqrt{2^t}}\sum_{j=0}^{2^t-1} e^{2\pi i\psi j}|j\rangle|u\rangle. \tag{11}$$

Applying the inverse QFT on the control register transforms the phase into a binary approximation $\overline{\psi}$. Measurement yields $\overline{\psi}$ with probability close to 1 provided the phase is close to a binary fraction. **Figure 3** illustrates the complete phase estimation circuit.

3.4 Probability of successful estimation

Let $\overline{\psi}$ be the n-bit approximation of $2^t\psi$. The measurement outcome is accurate to within δ if:

$$2^t\psi = \overline{\psi} + \delta, \quad \text{with} \quad 0 \le \delta < \frac{1}{2^{t+1}}. \tag{12}$$

The final measurement in the computational basis yields $\overline{\psi}$ with probability:

$$\Pr(\overline{\psi}) = \begin{cases} 1 & \text{if } \delta = 0, \\ \left| \frac{1 - e^{2\pi i 2^t \delta}}{2^t (1 - e^{2\pi i \delta})} \right|^2 & \text{if } \delta \neq 0. \end{cases} \tag{13}$$

This shows the phase estimation algorithm succeeds with high probability, especially as t increases.

3.5 Computational complexity

The phase estimation algorithm requires:

- $O(t^2)$quantum gates for inverse QFT,
- One application of the unitary U raised to powers of two,
- $t = n + [\log(2+1/2\epsilon)]$ qubits to achieve n-bit precision with error ϵ.

Detailed complexity analysis is presented in Nielsen and Chuang's foundational text [12].

4. Cryptographic vulnerabilities in the quantum era

The transition from classical to quantum computing has profound implications for the security assumptions that underpin modern cryptographic systems. While conventional cryptography relies on the presumed intractability of certain mathematical problems, quantum algorithms directly target these hardness assumptions, rendering once-secure schemes vulnerable. This section examines the most critical threats posed by quantum algorithms and their broader consequences for public-key infrastructures (PKIs) and digital security as a whole.

4.1 The threat of quantum algorithms

The central challenge posed by quantum computing to modern cryptography stems from its ability to address problems that remain infeasible for classical methods. The clearest illustration is Shor's algorithm, which shows that integer factorization and the discrete logarithm problem can both be resolved in polynomial time once a sufficiently powerful quantum computer is available. This breakthrough directly compromises the foundations of widely used public-key mechanisms, including RSA, Diffie–Hellman key exchange, and elliptic curve cryptography (ECC). To illustrate, RSA relies on the classical difficulty of factoring the product of two large primes, a task that requires sub-exponential resources with the fastest known conventional technique, the Number Field Sieve. Shor's approach, however, reduces this task to $\mathcal{O}(n^3)$ operations with $\mathcal{O}(n^2)$ memory, where n represents the modulus size in bits. Likewise, the elliptic curve discrete logarithm problem (ECDLP), which secures ECC, can also be efficiently broken under Shor's framework. As a result, many of the core primitives supporting digital security would fail once scalable quantum machines become available.

In addition to Shor's algorithm, Grover's algorithm introduces additional challenges, particularly for symmetric key cryptography. Although symmetric ciphers such as AES and hash functions remain more robust to quantum attacks than their public-key counterparts, Grover's algorithm provides a quadratic speedup for brute-force key search and preimage attacks. This effectively reduces the security strength of symmetric systems: a k-bit key offers only $k/2$ bits of quantum security. For example, AES-128, which provides 128-bit classical security, is reduced to 64-bit security against a quantum adversary, which is insufficient for long-term protection. To maintain the adequate post-quantum security levels, symmetric key sizes must therefore be doubled (e.g., AES-256). While this adjustment preserves the viability of symmetric encryption, it emphasizes the disruptive asymmetry of the quantum threat: Public-key primitives are fundamentally broken, while symmetric schemes require only parameter adjustments.

4.2 Collapse of traditional public-key infrastructure (PKI)

The vulnerabilities exposed by quantum algorithms extend far beyond encryption and key exchange. Public-key cryptography also underpins essential services such as digital signatures, authentication protocols, and certificate-based infrastructures. If adversaries equipped with quantum computers can efficiently forge digital signatures, impersonate entities, or retroactively decrypt stored ciphertexts, the trust model of the internet collapses. This scenario is often described as the "quantum apocalypse," the sudden obsolescence of the global PKI [19].

In practical terms, the collapse of PKI would disrupt secure web browsing (TLS/SSL), and secure email protocols (PGP, S/MIME), blockchain and cryptocurrency systems, and virtually all digital identity management systems. Certificates signed with RSA or ECC would no longer be trustworthy, and adversaries could both impersonate legitimate servers and decrypt historical communications that were recorded in encrypted form (a phenomenon known as "store now, decrypt later"). The risk is therefore not merely prospective but also retrospective, threatening the confidentiality of data already transmitted under classical schemes.

To mitigate these risks, a fundamental redesign of PKI is required, integrating quantum-safe primitives that remain secure against both classical and quantum adversaries. Post-Quantum Cryptography offers candidate algorithms, such as lattice-based schemes (e.g., Kyber, Dilithium), code-based encryption (e.g., Classic McEliece), and hash-based signatures (e.g., SPHINCS+), that can replace RSA and ECC while maintaining compatibility with the existing network protocols. Transitioning to such primitives is a complex socio-technical challenge, involving standardization (e.g., the NIST PQC project), software and hardware integration, and widespread migration across critical infrastructure. Nonetheless, this transition is essential to ensure the resilience of digital security in the quantum era.

5. Mathematical foundations of post-quantum cryptography

Post-Quantum Cryptography is grounded in mathematical problems that are believed to resist both classical and quantum attacks. Unlike RSA and ECC, which rely on factorization and discrete logarithms, PQC schemes are built upon problems for which no efficient quantum algorithms are known. This section surveys the main families of post-quantum cryptographic constructions and the mathematical assumptions that support their security.

5.1 Lattice-based cryptography

At the forefront of PQC research is lattice-based cryptography [20], widely regarded as the most versatile and efficient class of quantum-resistant systems. Lattices are discrete periodic structures in $\mathbb{R}^n$, and their complexity arises from the difficulty of finding short or close vectors within these grids. The most important hard problems include:

- *Shortest vector problem (SVP) and closest vector problem (CVP)*: Given a lattice basis, finding the shortest non-zero vector or the closest lattice vector to a target point is conjectured to remain hard even for quantum computers [21].
- *Learning with errors (LWE) and ring-LWE*: These problems introduce noise into linear equations over modular arithmetic, making recovery of the secret vector computationally infeasible [22]. They provide the security foundation for leading NIST candidates such as *CRYSTALS-Dilithium* (digital signatures) and *Kyber* (key encapsulation).

Lattice-based cryptography offers strong worst-case to average-case reductions, meaning that breaking one random instance is as hard as solving the hardest case of the underlying lattice problem [23]. Moreover, structured variants such as Ring-LWE allow efficient implementations, making these schemes well-suited for practical deployment.

5.2 Code-based cryptography

Code-based cryptography, first proposed by McEliece in 1978 [24], builds security on the hardness of decoding random linear codes. The fundamental assumption is the Syndrome Decoding Problem; given a parity-check matrix and a noisy codeword, recovering the original message is computationally hard. Prominent candidates in this category include *Classic McEliece*, which is highly resistant to quantum attacks, and *BIKE*, which leverages structured codes for efficiency. Code-based schemes benefit from decades of cryptanalytic scrutiny and remain strong contenders for long-term security. Their main drawback is the large size of public keys, though modern variants have improved compactness [25].

5.3 Multivariate polynomial cryptography

Multivariate cryptography constructs schemes from systems of quadratic equations over finite fields [26]. Solving such systems, known as the MQ problem, is NP-hard, providing the basis for security. These schemes are particularly attractive for digital signatures, with *Rainbow* being a notable NIST candidate [27]. The strengths of multivariate systems lie in their efficient signing operations and mathematical simplicity. However, they often suffer from very large public key sizes and occasional structural weaknesses discovered through algebraic cryptanalysis, which has led to the deprecation of some candidates during the NIST process.

5.4 Isogeny-based cryptography

Isogeny-based cryptography is the most recent family of PQC constructions, relying on the difficulty of computing isogenies (structure-preserving maps) between elliptic curves [28]. Supersingular isogeny graphs provide a rich algebraic structure believed to be resistant to both classical and quantum attacks.

- Supersingular Isogeny Diffie–Hellman (SIDH) [29]: Once considered a leading candidate, SIDH was broken by powerful algebraic attacks (Castryck–Decru, 2022).
- CSIDH and other variants [30]: These aim to retain the compact key sizes and unique algebraic features of isogeny-based schemes while improving resilience against cryptanalysis.

Despite recent setbacks, isogeny-based cryptography remains of research interest due to its potential for extremely small key sizes and novel algebraic foundations.

5.4.1 *Algebraic number theory in isogeny-based cryptography*

Isogeny-based cryptography exploits maps between elliptic curves defined over finite fields. An *isogeny* is a non-constant rational morphism $\varphi : E_1 \to E_2$ that preserves the group law. The security of schemes such as CSIDH and SQISign relies on the presumed hardness of computing paths in supersingular isogeny graphs. From a number theoretic perspective, these constructions are deeply tied to class groups of quadratic fields and quaternion algebras. For instance, the CSIDH protocol uses the action of the ideal class group of $\mathbb{Q}(\sqrt{-p})$ on elliptic curve isomorphism classes, embedding algebraic number theory directly into cryptographic key exchange. The unresolved problem of computing endomorphism rings efficiently for supersingular curves illustrates how open number theoretic questions directly translate into cryptographic assumptions.

5.5 Cyclotomic fields and ring-LWE

A key structural ingredient in lattice-based post-quantum cryptography is the use of *cyclotomic fields* and their associated rings of integers. Recall that the n-th cyclotomic polynomial is defined as

$$\Phi_n(x) = \prod_{\substack{1 \le k \le n \\ \gcd(k,n)=1}} \left(x - e^{2\pi i k/n}\right). \tag{14}$$

For n a power of two, $\Phi_n(x) = x^{n/2} + 1$, yielding highly structured rings

$$R_q = \mathbb{Z}_q[x]/(\Phi_n(x)), \tag{15}$$

which form the algebraic backbone of Ring-LWE and Module-LWE schemes. This structure enables efficient arithmetic *via* the Number Theoretic Transform (NTT), while simultaneously connecting security to deep problems in algebraic number theory. In particular, the hardness of Ring-LWE is linked to the difficulty of finding short vectors in ideal lattices associated with cyclotomic fields. This interplay between efficient computation and provable reductions highlights why cyclotomic fields are central to modern PQC.

6. Role of number theory in PQC

Modern PQC schemes continue to draw heavily from classical number theory and algebra. For example:

- Modular arithmetic underpins lattice- and code-based systems, where modular reductions and linear algebraic operations dominate computations [31].
- Algebraic number fields and quaternion algebras provide the structural backbone of isogeny-based schemes [28].
- Cyclotomic fields and polynomial rings are used to optimize structured lattice schemes, enabling efficient implementations of Ring-LWE and Module-LWE systems [32].

This interplay between algebra, geometry, and arithmetic highlights the enduring centrality of number theory in cryptographic design. The same mathematical traditions that enabled RSA and ECC now guide the construction of secure alternatives in the quantum era.

7. NIST standardization process and practical transition

7.1 Overview of the NIST PQC project

In 2016, the U.S. National Institute of Standards and Technology (NIST) initiated an open competition to identify and standardize quantum-resistant cryptographic primitives. The evaluation process emphasized [33]:

- *Provable security*, grounded in hard, well-studied mathematical problems.
- *Implementation efficiency*, ensuring suitability for a wide range of classical hardware platforms.
- *Resilience to side-channel attacks*, ensuring practical security in real-world systems.

7.2 Finalists and selected algorithms

In 2022, NIST announced its first selections for standardization. *Kyber* was chosen as the primary algorithm for public-key encryption and key encapsulation, while *CRYSTALS-Dilithium* was selected for digital signatures. Additional recommendations include *Falcon* (a lattice-based signature scheme) and *SPHINCS+* (a hash-based signature scheme). These selections reflect a balance between mathematical soundness, efficiency, and maturity under cryptanalysis.

7.3 Migration challenges

Transitioning from classical public-key systems to PQC presents significant challenges. Integration into protocols such as TLS and VPNs, and secure email requires careful consideration of:

- *Backward compatibility* with existing infrastructure, ensuring continuity of secure communications during migration.
- *Implementation audits*, including side-channel resistance, memory usage, and performance under constrained devices.

The migration to PQC will likely be gradual, with hybrid schemes (combining classical and quantum-resistant algorithms) deployed during the transition phase. Nevertheless, the urgency of preparing digital infrastructures for the quantum era cannot be overstated.

8. Open research problems and future directions

Despite rapid advances in PQC, several open questions remain:

- *Lattice problems:* Are there sub-exponential quantum algorithms for Ring-LWE or Module-LWE using hidden subgroup techniques?
- *Isogenies:* Can new class-group-based constructions avoid the vulnerabilities that broke SIDH, while retaining compact key sizes?
- *Code-based systems:* How can key sizes be further reduced without compromising security?
- *Multivariate systems:* Can we design trapdoor functions that resist algebraic cryptanalysis while remaining efficient?
- *Global adoption:* Beyond NIST, how will regions such as Africa approach PQC migration, especially in lightweight IoT and mobile environments?

These challenges underscore the continuing dialog between number theory, algebraic geometry, and cryptography. The frontier of PQC is therefore not only an engineering problem but also a fertile ground for pure mathematical research.

Toy problems serve as simplified illustrations of otherwise intractable cryptographic challenges, allowing readers to build intuition before engaging with the full mathematical complexity. For example, the Shortest Vector Problem (SVP) can be introduced using a two-dimensional lattice generated by simple basis vectors, where identifying the shortest vector is straightforward, yet the same task becomes computationally infeasible in high dimensions. Similarly, the syndrome decoding problem may be demonstrated with a small parity-check matrix and a single error, showing how easy cases contrast with the hardness underlying code-based cryptography. Even in quantum algorithms, period finding for small modular functions (e.g., $f(x) = 2^x \text{mod} 15$) offers a concrete stepping stone toward understanding Shor's algorithm. These illustrative exercises not only demystify abstract problems but also highlight why their large-scale versions provide the foundation for post-quantum security.

9. Conclusion

The advent of large-scale quantum computing poses an unprecedented challenge to the security foundations of modern cryptography. The Shor and Grover algorithms demonstrate that RSA, ECC, and even symmetric primitives cannot remain unaltered in the quantum era. Post-quantum cryptography offers a spectrum of alternatives, lattice-based, code-based, multivariate, and isogeny-based schemes grounded in hard mathematical problems that resist both classical and quantum attacks. A recurring theme throughout this chapter is the enduring role of number theory, which not only shaped the foundations of RSA and ECC but also guides the construction of quantum-resistant protocols. The NIST standardization process has accelerated the transition from theoretical constructs to practical deployment, with algorithms like Kyber and Dilithium now positioned as future cornerstones of global security infrastructures. However, migration to PQC presents both technical and organizational challenges, from ensuring backward compatibility to defending against implementation-level vulnerabilities. As research continues, hybrid models and further cryptanalytic scrutiny will be vital in guaranteeing the long-term robustness of quantum-safe primitives. Ultimately, the future of secure communication will rely on the successful integration of these post-quantum tools, ensuring resilience in the face of quantum adversaries.

Acknowledgements

The authors acknowledge the Council for Scientific and Industrial Research (CSIR) and the Department of Science and Innovation (DSI) for the grant of funding for this research. K.M. was also supported by the South African Quantum Technology Initiative (SAQuTi) and South African Medical Research Council (SAMRC).

Conflict of interest

The authors declare no conflict of interest.

Nomenclature

AES	advanced encryption standard
CVP	closest vector problem
ECDLP	elliptic curve discrete logarithm problem
ECC	elliptic curve cryptography
KEM	key encapsulation mechanism
LWE	learning with errors
NIST	National Institute of Standards and Technology
NTRU	N-th degree truncated polynomial ring units
PKI	public-key infrastructure
PQC	post-quantum cryptography
QFT	quantum Fourier transform
RSA	Rivest–Shamir–Adleman cryptosystem
SIDH	supersingular isogeny Diffie–Hellman
SPHINCS+	stateless practical Hash-based incredibly nice cryptographic signature
SVP	shortest vector problem

Post-Quantum Cryptography: Number Theoretic Foundations and Future-Proof Protocols
DOI: http://dx.doi.org/10.5772/intechopen.1012847

Author details

Kelvin Tafadzwa Mpofu[1*†] and Patience Mthunzi-Kufa[1,2,3†]

1 Biophotonics, Photonic Centre, Manufacturing, Council of Scientific and Industrial Research (CSIR), Pretoria, South Africa

2 Molecular and Cell Biology Department, University of Cape Town, South Africa

3 School of Interdisciplinary Research and Graduate Studies (UNESCO), College of Graduate Studies, University of South Africa, Pretoria, South Africa

*Address all correspondence to: kmpofu@csir.co.za

† These authors contributed equally.

IntechOpen

© 2025 The Author(s). Licensee IntechOpen. This chapter is distributed under the terms of the Creative Commons Attribution License (http://creativecommons.org/licenses/by/4.0), which permits unrestricted use, distribution, and reproduction in any medium, provided the original work is properly cited.

References

[1] Pachghare V. Cryptography and Information Security. 3rd ed. New Delhi, India: PHI Learning Pvt. Ltd.; 2019

[2] Halpin H. The adversary: The philosophy of cryptography. Journal of Cybersecurity. 2025;**11**(1):tyaf006

[3] Easttom C. Modern Cryptography: Applied Mathematics for Encryption and Information Security. Cham, Switzerland: Springer; 2015

[4] Delfs H, Knebl H. Symmetric-key cryptography. In: Introduction to Cryptography: Principles and Applications. 3rd ed. Berlin, Heidelberg: Springer; 2015. pp. 11-48

[5] Merkle RC. Secure communications over insecure channels. Communications of the ACM. 1978;**21**(4):294-299

[6] Saho NJG, Ezin EC. Comparative study on the performance of elliptic curve cryptography algorithms with cryptography through rsa algorithm. In: Watson B, Badouel E, Niang O, editors. Proceedings of CARI 2020 – Colloque Africain sur la Recherche en Informatique et en Mathématiques Appliquées. Thiés, Senegal: École Polytechnique de Thiés, October 2020, hAL Id: hal-02926106. [Online]. Available from: https://hal.science/hal-02926106

[7] Koblitz N. A Course in Number Theory and Cryptography, 2nd ed., ser. Graduate Texts in Mathematics. Vol. 114. New York, NY, USA: Springer; 1994

[8] Abdullah K. Comparison between the Rsa Cryptosystem and Elliptic Curve Cryptography [Ph.D. dissertation]. Hamilton, New Zealand: The University of Waikato; 2010

[9] Gisin N, Ribordy G, Tittel W, Zbinden H. Quantum cryptography. Reviews of Modern Physics. 2002;**74**(1): 145

[10] Portmann C, Renner R. Security in quantum cryptography. Reviews of Modern Physics. 2022;**94**(2):025008

[11] Shor PW. Algorithms for quantum computation: Discrete logarithms and factoring. In: Proceedings 35th Annual Symposium on Foundations of Computer Science (FOCS). Los Alamitos, CA, USA: IEEE Computer Society; 1994. pp. 124-134

[12] Nielsen MA, Chuang IL. Quantum Computation and Quantum Information. 10th ed. Cambridge, United Kingdom: Cambridge University Press; 2010

[13] Grover LK. A fast quantum mechanical algorithm for database search. In: Proceedings of the Twenty-Eighth Annual ACM Symposium on Theory of Computing (STOC). New York, NY, USA: Association for Computing Machinery; 1996. pp. 212-219

[14] Bagirovs E, Provodin G, Sipola T, Hautamäki J. Applications of post-quantum cryptography. Vol. 1. arXiv preprint arXiv:2406.13258. 2024. pp. 1-15. [Online]. Available: https://arxiv.org/abs/2406.13258. 2024

[15] Sahu SK, Mazumdar K. State-of-the-art analysis of quantum cryptography: Applications and future prospects. Frontiers in Physics. 2024;**12**:1456491

[16] Moody D. Nist PQC Standardization Update. In: National Institute of Standards and Technology (NIST), 2021, presented at the NIST PQC Standardization Conference. Oct 2021. [Online]. Available from: https://csrc.nist.gov/Presentations/2021/pqc-standardization-update-2021

[17] Bavdekar R, Chopde EJ, Agrawal A, Bhatia A, Tiwari K. Post quantum cryptography: A review of techniques, challenges and standardizations. In: 2023 International Conference on Information Networking (ICOIN). Bangkok, Thailand: IEEE; 2023. pp. 146-151

[18] Steane A. Quantum computing. Reports on Progress in Physics. 1998; **61**(2):117

[19] Bene F, Kiss A. Public key infrastructure in the post-quantum era. In: 2023 IEEE 17th International Symposium on Applied Computational Intelligence and Informatics (SACI). Timisoara, Romania: IEEE; 2023. pp. 000077-000082

[20] Nejatollahi H, Dutt N, Ray S, Regazzoni F, Banerjee I, Cammarota R. Post-quantum lattice-based cryptography implementations: A survey. ACM Computing Surveys (CSUR). 2019;**51**(6):1-41

[21] Voulgaris P. Algorithms for the Closest and Shortest Vector Problems on General Lattices [Ph.D. Thesis]. San Diego, La Jolla, CA, USA: University of California; 2011

[22] Lyubashevsky V, Peikert C, Regev O. On ideal lattices and learning with errors over rings. In: In Advances in Cryptology – EUROCRYPT 2010, ser. Lecture Notes in Computer Science. Vol. 6110. Berlin, Heidelberg: Springer; 2010. pp. 1-23

[23] Zhang J, Zhang Z. Lattice-Based Cryptosystems, ser SpringerBriefs in Computer Science. Singapore: Springer; 2020

[24] McEliece RJ. A public-key cryptosystem based on algebraic. Coding Thv. 1978;**4244**(1978):114-116

[25] Nguyen V. Code-based cryptography: Attacking and constructing cryptographic systems. [Doctoral thesis]. Lund, Sweden: Lund University; May 2025. series of Licentiate and Doctoral Theses No. 186, ISSN 1654-790X-186

[26] Wolf C. Multivariate quadratic polynomials in public key cryptography. In: Cryptology ePrint Archive. Vol. 1. 2005. pp. 1-20. [Online]. Available from: https://eprint.iacr.org/2005/393

[27] Shajahan R, Jain K, Krishnan P. A survey on nist 3 rd round post quantum digital signature algorithms. In: 2024 5th International Conference on Mobile Computing and Sustainable Informatics (ICMCSI). Pune, India: IEEE; 2024. pp. 132-140

[28] De Feo L. Mathematics of isogeny based cryptography. arXiv preprint arXiv:1711.04062. 2017. pp. 1-35

[29] Costello C, Longa P, Naehrig M. Efficient algorithms for supersingular isogeny diffie-hellman. In: Annual International Cryptology Conference. Berlin, Heidelberg: Springer; 2016. pp. 572-601

[30] Kutas P, Martindale C, Panny L, Petit C, Stange KE. Weak instances of sidh variants under improved torsion-point attacks. IACR Cryptology ePrint Archive. 2020;**633**:1-20

[31] Ganesh MAD. Recent advances and innovations in algebraic structures and their applications. Mathematical Innovation. 2025;**1**:1

[32] Washington LC. Introduction to Cyclotomic Fields. Vol. 83. New York, NY, USA: Springer Science and Business Media; 2012

[33] Plan NPA. National institute of standards and technology (nist)

Chapter 6

Ghani Mersenne and Fermat Entropy for Some Dendrimers and Their Structure-Property Modeling

Muhammad Usman Ghani, Syed Ajaz K. Kirmani, Faisal Sultan and Fairouz Tchier

Abstract

The mathematical language provided by various topological index types may be used to identify various properties of chemical components in a molecular structure. Within the framework of valency-based topological indices, we examine eight different dendrimer architectures. This chapter proposes two novel molecular descriptors, the Mersenne index and the Fermat index, by combining the essential notions of the Mersenne number and the Fermat number. The molecules with the highest Mersenne index and Fermat index values are found in the classes of some dendrimer families, including poly or propyl ether-imine dendrimer, polyethylene amido amine dendrimer, porphyrin dendrimer, zinc porphyrin dendrimer, porphyrin-cored dendrimer-I, porphyrin-cored dendrimer-II, porphyrin-cored dendrimer-III, and porphyrin-cored dendrimer-IV. The Fermat index differs from the Mersenne index in that it grows with the value of n, as the Fermat number is always increasing by increasing the value of n. Also, we show that the Mersenne index and the Fermat index are always members of the set of integers (Z) for any graph of chemical compounds.

Keywords: Mersenne index, Fermat index, topological indices, entropies, porphyrin-cored dendrimers

1. Introduction

Chemical graph theory is a branch of graph theory in which the molecular structure is made up of atoms and atom bonds, which are represented by the graph's vertices and edges, respectively [1]. To forecast the chemical structure of a graph, as well as graph theory, aids in determining the structural properties of a graph. Similarly, topological indices have numerous applications in theoretical chemistry, such as in the field of quantum chemistry of QSPR/QSAR studies, see Ref. [2, 3]. The molecular graphs in chemical graph theory are simple connected graphs that contain chemical atoms and bonds, which are often known as vertices and edges [4], respectively, and there must be links among the vertices set V_G and edges set E_G. The degree of each vertex of G is actually the total number of vertices linked to v of G and it is denoted by d_{v_i} [5, 6].

Some of the previously defined topological indices give the values in points; some indices [7, 8] do not provide precise values for their indices; rather, they always provide an approximation, like the Randic index $\left(R(G) = \frac{1}{\sqrt{d_u \times d_v}}\right)$, the atom bond index $\left(ABC = \sqrt{\frac{d_u+d_v-2}{d_u \times d_v}}\right)$, the atom-bond sum connectivity index [5], $\left(ABC = \sqrt{\frac{d_u+d_v-2}{d_u \times d_v}}\right)$, the geometric index and the fourth version of the ABC index, etc [9–12]. All these indices and other indices that have square roots in their formulae always give us an approximate value, and their equations are not in polynomial form [7, 12]. We cannot compute exact values from these indices for a molecular structure. We need to define new indices that can give us exact values or values from a set of integers (Z) [13–15].

To solve these problems, we defined the Ghani Mersenne polynomial and Ghani Fermat polynomial; with the help of these polynomials, we can compute their related indices [16–19]. Taking the derivative at $x = 1$ of these polynomials, we get the Ghani Mersenne index and the Ghani Fermat index [20–24]. These indices always give a positive integer [25–27]. The search for Mersenne primes is a popular topic in number theory and computer science. It is also one of the most important applications for distributed computing, which is a technique in which hundreds of computers are linked over the Internet and work together to solve a problem [28–33].

The numbers with the form $M_n = 2^n - 1$, where n is a positive integer, are known as Mersenne numbers. The Ghani Mersenne polynomial and Ghani Mersenne index were defined as follows using the definition of the Mersenne number:

$$G_{\mathrm{m}}(x) = \sum_{v_i \in V(G)} x^{\left(2^{d_{v_i}}-1\right)} \quad \text{and} \quad G_{\mathrm{m}} = \sum_{v_i \in V(G)} \left(2^{d_{v_i}} - 1\right) \tag{1}$$

In mathematics, a Fermat number, named after Pierre de Fermat, who first studied them, is a positive integer of the form:

$$F_n = 2^{2^n} + 1$$

where n is a non-negative integer. The Fermat Ghani polynomial and Fermat Ghani index is given below:

$$G_{\mathrm{f}}(x) = \sum_{v_i \in V(G)} x^{\left(2^{2^{d_{v_i}}}-1\right)} \quad \text{and} \quad G_{\mathrm{f}} = \sum_{v_i \in V(G)} \left(2^{2^{d_{v_i}}} - 1\right) \tag{2}$$

2. Poly or propyl ether-imine dendrimer

The poly or propyl ether-imine, denoted by *PETIM*, is a dendrimer of generation G_n, where n denotes its growth stages, for $n \geq 1$. *PETIM* contains one central core and four branches each branch is attached with n by atom-bond $\sim$; see **Figure 1**.

2.1 Results and discussion

As illustrated in **Figure 1**, the center core comprises 7 atoms of valency 2, 2 atoms of valency 3, and 8 atom bonds ($\sim$). In one branch of *PETIM*, there are total $2(2^n - 4)$ atoms, in which 2^{n-1} atoms of valency 1, 2^{n+1} atoms of valency 3 and $7(2^{n-1} - 1)$

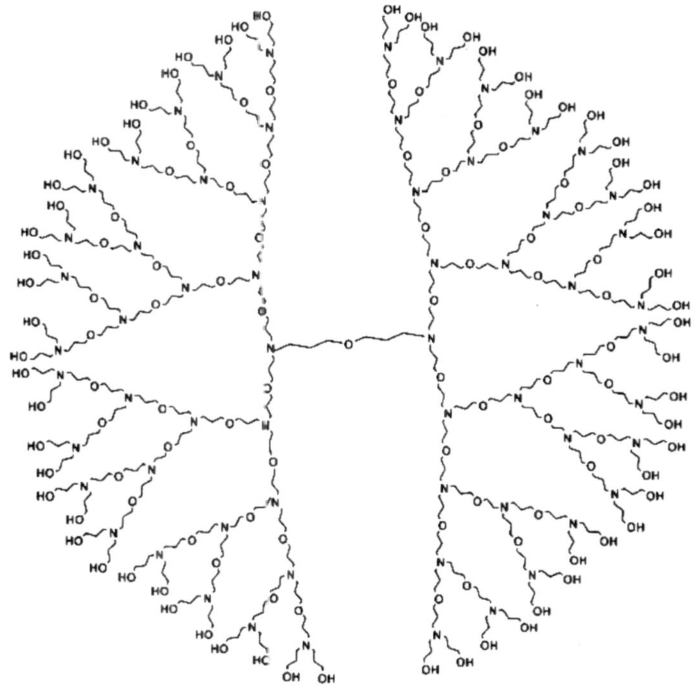

Figure 1.
Poly or propyl ether-imine dendrimer.

Type of atoms	$d_{v_1} = 1$	$d_{v_2} = 2$	$d_{v_3} = 3$
Number of atoms	2.2^n	$20.2^n - 21$	$2(2^n - 1)$

Table 1.
Types of atom in poly or propyl ether-imine deadrimer.

atoms of valency 2. As a result of the helt of our observation, we obtain **Table 1**. As a result of **Table 1**, *PETIM* contains $3.2^{n+3} - 23$ atoms.

We get these polynomials by using **Table 1** (types of atoms in poly or propyl ether-imine dendrimer) in the formulae of Ghani Mersenne polynomial in Eq. (1) and Ghani Fermat polynomial in Eq. (2).

$$G_{\mathrm{m}}(PETIM, x) = \sum_{\forall v_i \in A(PETIM)} x^{\left(2^{d_{v_i}} - 1\right)} \tag{3}$$

$$= \sum_{v_1 \in A(PETIM)} x^{\left(2^1 - 1\right)} + \sum_{v_2 \in A(PETIM)} x^{\left(2^2 - 1\right)} + \sum_{v_3 \in A(PETIM)} x^{\left(2^3 - 1\right)} \tag{4}$$

$$= 2.2^n x + (20.2^n - 21)x^3 + 2(2^n - 1)x^7. \tag{5}$$

$$G_{\mathrm{f}}(PETIM, x) = \sum_{v_i \in A(PETIM)} x^{\left(2^{2^{d_{v_i}}} - 1\right)} \tag{6}$$

Values of n	1	2	3	4	5	6	7	8	9	10
G_m	79	235	547	1171	2419	4915	9907	19,891	39,859	79,795
G_f	1062	2694	5958	12,486	25,542	51,654	103,878	208,326	417,222	835,014

Table 2.
Numerical comparison of Ghani Mersenne and Ghani Fermat indices of PETIM.

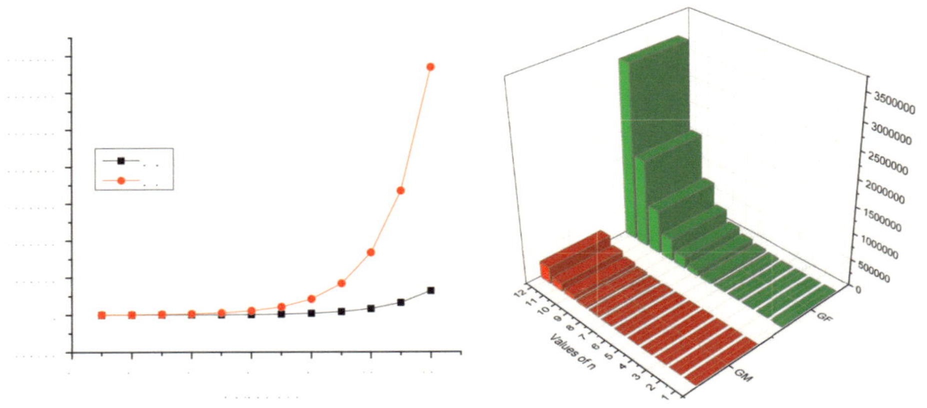

Figure 2.
Graphical comparison in 2D and 3D graphs.

$$= \sum_{v_1 \in A(PETIM)} x^{\left(2^{2^1}-1\right)} + \sum_{v_2 \in A(PETIM)} x^{\left(2^{2^2}-1\right)} + \sum_{v_3 \in A(PETIM)} x^{\left(2^{2^3}-1\right)} \quad (7)$$

$$= 2.2^n x^3 + (20.2^n - 21)x^{15} + 2(2^n - 1)x^{255}. \quad (8)$$

Taking the derivative of each polynomial from (4) and (7) at $x = 1$ yields the Ghani Mersenne topological index and the Ghani Fermat topological index:

$$T_i(PETIM) = \begin{cases} 76.2^n - 77, & \text{whenever } T_i = G_m, \\ 2(408.2^n - 285), & \text{whenever } T_i = G_f. \end{cases}$$

2.2 Numerical comparison of indices of *PETIM*

In this section, we present a numerical and graphical comparison of the Ghani Mersenne topological index and the Ghani Fermat topological index for $n = 1, 2, 3, 4, \ldots, 10$ of *PETIM* in **Table 2** and **Figure 2**.

3. Poly ethylene amido amine dendrimer

The poly ethylene amido-amine (*PETAA*) dendrimer, of generation G_n, where n denotes its growth stages for $n \geq 1$, is our second molecular family. The molecule *PETAA* comprises four branches and one central core, see **Figure 3**.

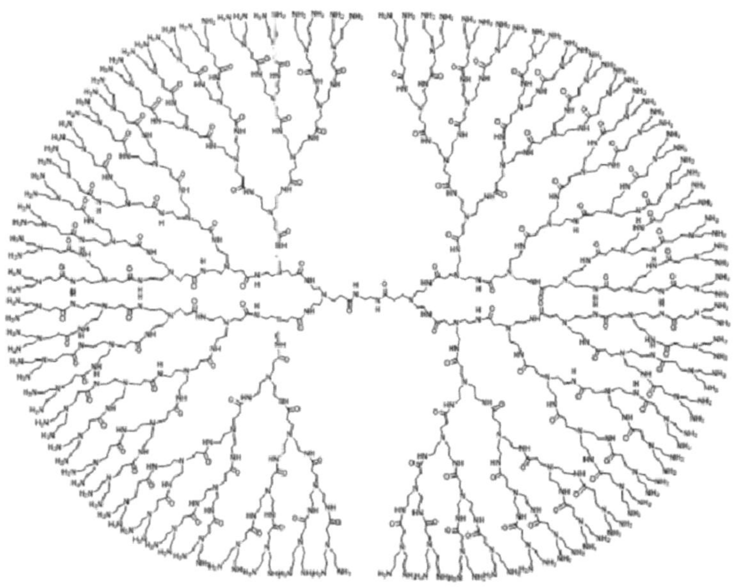

Figure 3.
Poly ethylene amido amine dendrimer.

Type of atoms	$d_{v_1} = 1$	$d_{v_2} = 2$	$d_{v_3} = 3$
Number of atoms	$2(4.2^n - 1)$	$4(7.2^n - 3)$	$4(2.2^n - 1)$

Table 3.
Types of atom in porphyrin-cored dendrimers-1.

3.1 Results and discussion

The center core of *PETAA* is shown in **Figure 3**. It has 13 atom-bonds, 8 valency 2 atoms, 4 valency 3 atoms, and 2 valency 1 atoms. In each branch, there are a total of $11.2^n - 8$ atoms, in which $2^{n+1} - 1$ atoms contain valency 1, $7.2^n - 5$ atoms contain valency 2 and $2^{n+1} - 2$ atoms contain valency 3. From these observations, we get **Table 3**. As a result of **Table 3**, *PETAA* contains $3.2^{n+3} - 23$ atoms.

We get these polynomials by using **Table 3** (types of atoms in *PETAA*) in the formulae of Ghani Mersenne polynomial in Eq. (1) and Ghani Fermat polynomial in Eq. (2).

$$G_m(PETAA, x) = \sum_{\forall v_i \in A(PETAA)} x^{\left(2^{d_{v_i}} - 1\right)} \tag{9}$$

$$= \sum_{v_1 \in A(PETAA)} x^{\left(2^1 - 1\right)} + \sum_{v_2 \in A(PETAA)} x^{\left(2^2 - 1\right)} + \sum_{v_3 \in A(PETAA)} x^{\left(2^3 - 1\right)} \tag{10}$$

$$= 2(4.2^n - 1)x + 4(7.2^n - 3)x^3 + 4(2.2^n - 1)x^7. \tag{11}$$

$$G_f(PETAA, x) = \sum_{\forall v_i \in A(PETAA)} x^{\left(2^{d_{v_i}} - 1\right)} \tag{12}$$

Values of n	1	2	3	4	5	6	7	8	9	10
G_m	230	526	1118	2302	4670	9406	18,878	37,822	75,710	151,486
G_f	3762	8730	18,666	38,538	78,282	157,770	316,746	634,698	1,270,602	2,542,410

Table 4.
Numerical comparison of Ghani Mersenne and Ghani Fermat indices of PETAA.

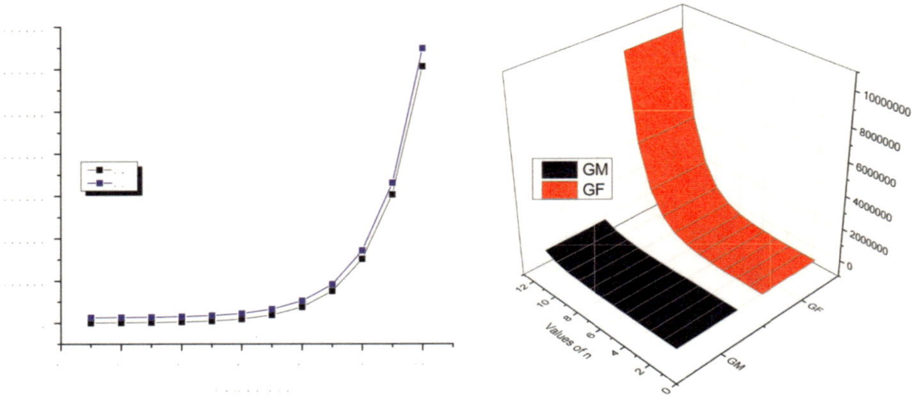

Figure 4.
Graphical comparison.

$$= \sum_{v_1 \in A(PETAA)} x^{\left(2^{2^1}-1\right)} + \sum_{v_2 \in A(PETAA)} x^{\left(2^{2^2}-1\right)} + \sum_{v_3 \in A(PETAA)} x^{\left(2^{2^3}-1\right)} \tag{13}$$

$$= 2(4.2^n - 1)x^3 + 4(7.2^n - 3)x^{15} + 4(2.2^n - 1)x^{255}. \tag{14}$$

Taking the derivative of each polynomial from (10) and (13) at $x = 1$ yields the Ghani Mersenne topological index and the Ghani Fermat topological index.

$$T_i(PETAA) = \begin{cases} 2(74.2^n - 33), & \text{whenever } T_i = G_m, \\ 18(138.2^n - 67), & \text{whenever } T_i = G_f. \end{cases}$$

3.2 Numerical comparison of indices of *PETAA*

This section compares, for $n = 1, 2, 3, \ldots, 10$ of *PETAA*, the Ghani Mersenne topological index with the Ghani Fermat topological index numerically and graphically in **Table 4** and **Figure 4**.

4. Porphyrin dendrimer

The porphyrin dendrimer, denoted by *PD*, of generation G_n, where n denotes its growth stages for $n \geq 2$, is our third molecular family. The molecule *PD* is made up of four branches and one central core, see in **Figure 5**.

4.1 Results and discussion

Figure 5 of *PD* depicts the central core, which has 4 atoms of valency 2, 2 atoms of valency 3, and 5 atom-bonds. In each branch, there are a total of $96.2^{n-2} - 4$ atoms, in which $26.2^{n} - 10$ atoms contain valency 1, $34.2^{n-2} - 3$ atoms contain valency 2, $28.2^{n-2} - 1$ atoms contain valency 3, and 8.2^{n-2} atoms contain valency 4. From these observations, we get **Table 5**.

$$G_m(PD, x) = \sum_{\forall v_i \in A(PD)} x^{\left(2^{d_{v_i}} - 1\right)} \tag{15}$$

$$= \sum_{v_1 \in A(PD)} x^{\left(2^1 - 1\right)} + \sum_{v_2 \in A(PD)} x^{\left(2^2 - 1\right)} + \sum_{v_3 \in A(PD)} x^{\left(2^3 - 1\right)} + \sum_{v_4 \in A(PD)} x^{\left(2^4 - 1\right)} \tag{16}$$

$$= \left(13.2^{n+1}\right)x + \left(17.2^{n+1} - 8\right)x^3 + 2\left(7.2^{n+1} - 1\right)x^7 + 4.2^{n+1}x^{15}. \tag{17}$$

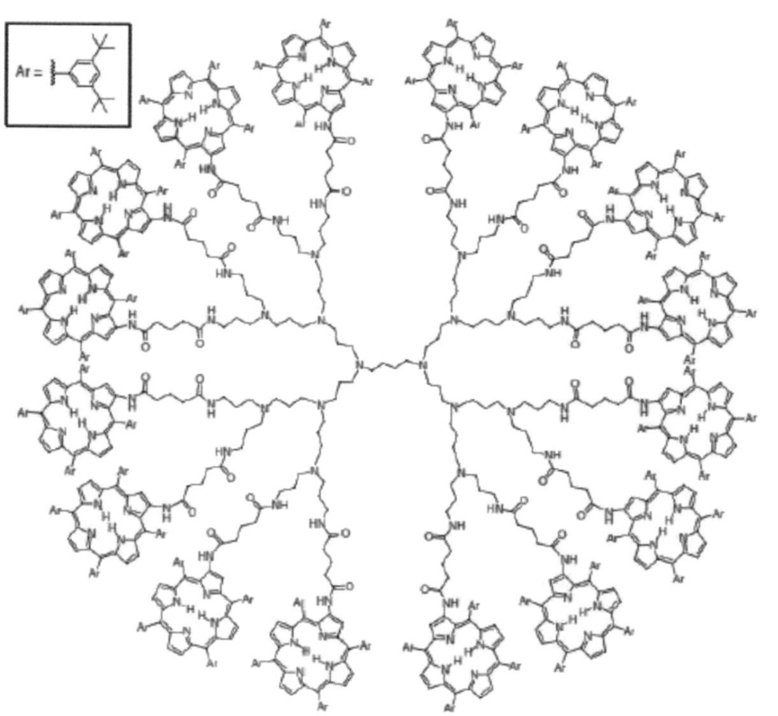

Figure 5.
Porphyrin dendrimer.

Type of atoms	$d_{v_1} = 1$	$d_{v_2} = 2$	$d_{v_3} = 3$	$d_{v_4} = 4$
Number of atoms	13.2^{n+1}	$(17.2^{n+1} - 8)$	$2(7.2^{n+1} - 1)$	4.2^{n+1}

Table 5.
Types of atom in porphyrin dendrimers.

$$G_{\mathfrak{f}}(PD, x) = \sum_{v_i \in A(PD)} x^{\left(2^{2^{d_{v_i}}} - 1\right)} \tag{18}$$

$$= \sum_{v_1 \in A(PD)} x^{\left(2^{2^1}-1\right)} + \sum_{v_2 \in A(PD)} x^{\left(2^{2^2}-1\right)} + \sum_{v_3 \in A(PD)} x^{\left(2^{2^3}-1\right)} + \sum_{v_4 \in A(PD)} x^{\left(2^{2^4}-1\right)} \tag{19}$$

$$= \left(13.2^{n+1}\right)x^3 + \left(17.2^{n+1} - 8\right)x^{15} + 2\left(7.2^{n+1} - 1\right)x^{255} + 4.2^{n+1}x^{65535}. \tag{20}$$

Taking the derivative of each polynomial from (16) and (19) at $x = 1$ yields the Ghani Mersenne topological index and the Ghani Fermat topological index.

$$T_i(PD) = \begin{cases} 2\left(111.2^{n+1} - 19\right), & \text{whenever } T_i = G_{\mathrm{m}}, \\ 18\left(14778.2^{n+1} - 35\right), & \text{whenever } T_i = G_{\mathfrak{f}}. \end{cases}$$

4.2 Numerical comparison of indices of *PD*

This section compares, for $n = 1, 2, 3, \ldots, 10$ of *PD*, the Ghani Mersenne topological index with the Ghani Fermat topological index numerically and graphically in **Table 6** and **Figure 6**.

5. Zinc porphyrin dendrimer

The zinc porphyrin dendrimer, denoted by *ZPD*, of generation G_n, where n denotes its growth stages for $n \geq 1$, is our fourth molecular family. The molecule *ZPD* comprises four branches and one central core, see **Figure 7**.

Values of n	G_{m}	$G_{\mathfrak{f}}$
1	850	1,063,386
2	1738	2,127,402
3	3514	4,255,434
4	7066	8,511,498
5	14,170	17,023,626
6	28,378	34,047,882
7	56,794	68,096,394
8	113,626	136,193,418
9	227,290	272,387,466
10	454,618	544,775,562
11	909,274	1,089,551,754
12	1,818,586	2,179,104,138

Table 6.
Numerical comparison of Ghani Mersenne and Ghani Fermat indices of PD.

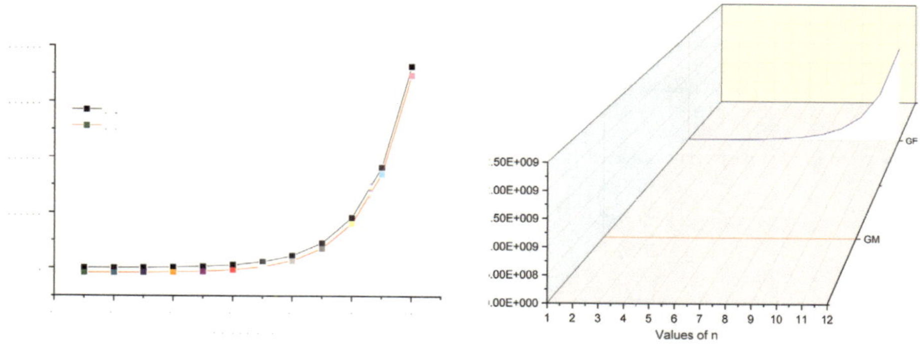

Figure 6.
Graphical comparison.

Figure 7.
Porphyrin dendrimer.

5.1 Results and discussion

Figure 7 of *ZPD* depicts the central core, which has 24 atoms of valencies 2 and 3, 1 atom of valency 4, and 60 atom bonds. In each branch, there are a total of 14.2^{n-1} atoms, in which $20.2^{n-1} - 9$ atoms contain valency 2, $4.2^{n} - 5$ atoms contain valency 3. From these observations, we get **Table 7**.

$$G_{\mathrm{m}}(ZPD, x) = \sum_{\forall v_i \in A(ZPD)} x^{\left(2^{d_{v_i}} - 1\right)} \tag{21}$$

Type of atoms	$d_{v_1}=2$	$d_{v_2}=3$	$d_{v_3}=4$
Number of atoms	$4(5.2^{n+1}-3)$	$4(2^{n+2}+1)$	2^2

Table 7.
Types of atom in zinc porphyrin dendrimer.

$$= \sum_{v_1 \in A(ZPD)} x^{(2^2-1)} + \sum_{v_2 \in A(ZPD)} x^{(2^3-1)} + \sum_{v_3 \in A(ZPD)} x^{(2^4-1)} \tag{22}$$

$$= 4(5.2^{n+1}-3)x^3 + 4(2^{n+2}+1)x^7 + 2^2x^{15}. \tag{23}$$

$$G_{\mathfrak{f}}(ZPD, x) = \sum_{v_i \in A(ZPD)} x^{\left(2^{2^{d_{v_i}}}-1\right)} \tag{24}$$

$$= \sum_{v_1 \in A(ZPD)} x^{\left(2^{2^2}-1\right)} + \sum_{v_2 \in A(ZPD)} x^{\left(2^{2^3}-1\right)} + \sum_{v_3 \in A(ZPD)} x^{\left(2^{2^4}-1\right)} \tag{25}$$

$$= 4(5.2^{n+1}-3)x^{15} + 4(2^{n+2}+1)x^{255} + 2^2x^{65535}. \tag{26}$$

Taking the derivative of each polynomial from (22) and (25) at $x = 1$ yields the Ghani Mersenne topological index and the Ghani Fermat topological index:

$$T_i(ZPD) = \begin{cases} 4(29.2^{n+1}+13), & \text{whenever } T_i = G_{\mathrm{m}}, \\ 20(117.2^{n+1}+13149), & \text{whenever } T_i = G_{\mathfrak{f}}. \end{cases}$$

5.2 Numerical comparison of indices of *PD*

This section compares, for $n = 1; 2, 3, 4, \dots, 10$ of *PD*, the Ghani Mersenne topological index with the Ghani Fermat topological index numerically and graphically in **Table 6** and **Figure 6**.

6. Porphyrin-cored dendrimers-I

The porphyrin-cored dendrimer-I, denoted by (PCD_I), is a single dendrimer molecule development stage in generation G_n with n indicating its stages. Porphyrin-cored dendrimer-I (PCD_I) is made up of two parallel branches and one central core, see **Figure 8**.

6.1 Results and discussion

As shown in **Figure 8**, the central core of PCD_I has 34 atom-bonds, 10 atoms of valency 2, 14 atoms of valency 3, and 1 atom of valency 4. Each branch of PCD_I contains a total of $2.2^n - 1$ atoms, with 2^n atoms having valency 1, no atoms of valency 2, no atoms of valency 3, and no atoms of valency 4. From these observations, we get **Table 8**.

$$G_{\mathrm{m}}(PCD_I, x) = \sum_{\forall v_i \in A(PCD_I)} x^{(2^{d_{v_i}}-1)} \tag{27}$$

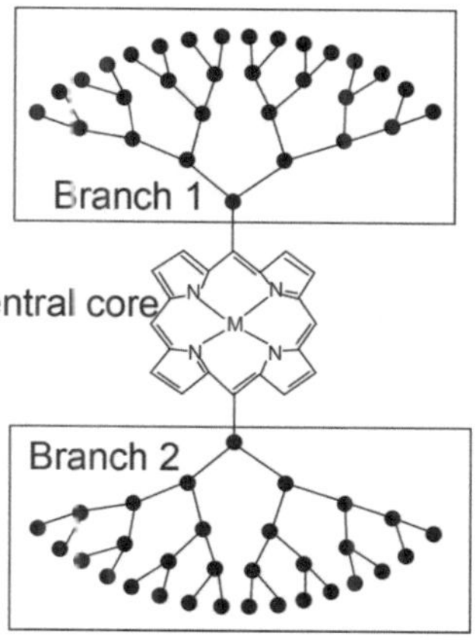

Figure 8.
Porphyrin-cored dendrimers-I.

Type of atoms	$d_{v_1}=1$	$d_{v_2}=2$	$d_{v_3}=3$	$d_{v_4}=4$
Number of atoms	2^{n-1}	10	$(2^{n+1}+12)$	1

Table 8.
Types of atom in porphyrin-cored dendrimers-I.

$$= \sum_{v_1 \in A(PCD_I)} x^{(2^1-1)} + \sum_{v_2 \in A(PCD_I)} x^{(2^2-1)} + \sum_{v_3 \in A(PCD_I)} x^{(2^3-1)} + \sum_{v_4 \in A(PCD_I))} x^{(2^4-1)} \tag{28}$$

$$= 2^{n+1}x + 10x^3 + (2^{n+1}+12)x^7 + x^{15}. \tag{29}$$

$$G_{\mathfrak{f}}(PCD_I, x) = \sum_{v_i \in A(PCD_I)} x^{\left(2^{2^{d_{v_i}}}-1\right)} \tag{30}$$

$$= \sum_{v_1 \in A(PCD_I)} x^{\left(2^{2^1}-1\right)} + \sum_{v_2 \in A(PCD_I)} x^{\left(2^{2^2}-1\right)} + \sum_{v_3 \in A(PCD_I)} x^{\left(2^{2^3}-1\right)} + \sum_{v_4 \in A(PCD_I)} x^{\left(2^{2^4}-1\right)} \tag{31}$$

$$= 2^{n+1}x^3 + 10x^{15} + (2^{n+1}+12)x^{255} + x^{65536}. \tag{32}$$

Taking the derivative of each polynomial from (28) and (31) at $x = 1$ yields the Ghani Mersenne topological index and the Ghani Fermat topological index.

$$T_i(PCD_I) = \begin{cases} 8.2^{n+1} + 129, & \text{whenever } T_i = G_{\mathrm{m}}, \\ 2(129.2^{n+1} + 34323), & \text{whenever } T_i = G_{\mathfrak{f}}. \end{cases}$$

7. Porphyrin-cored dendrimer-II

The development stage of the generation G_n porphyrin-cored dendrimer-II (PCD_{II}) is $n \geq 1$. As seen in **Figure 9**, PCD_{II} is made up of a single central core and four

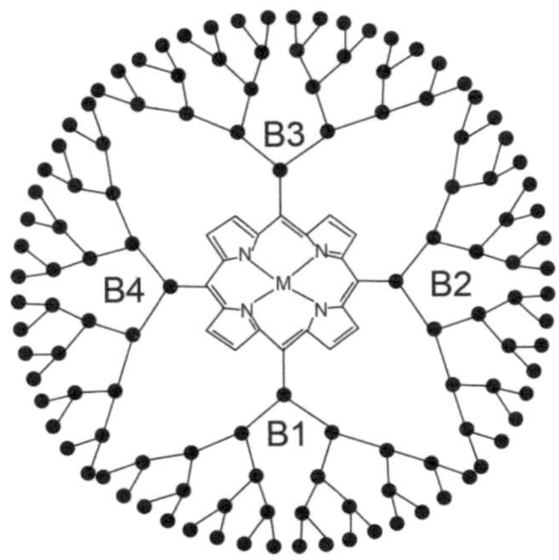

Figure 9.
Porphyrin-cored dendrimer-II.

parallel branches. The central core is made up of 36 atom-bonds, 1 atom of valency 4, 16 atoms of valency 3, and 8 atoms of valency 2, as shown in **Figure 9**.

7.1 Results and discussion

There are a total of $2.2^n - 1$ atoms in each branch of (PCD_{II}), with 2^n atoms of valency 1, zero atoms of valency 2, $2^n - 1$ atoms of valency 3, and zero atoms of valency 4. At the conclusion, we get **Table 9**.

$$G_{\mathrm{m}}(PCD_{II}, x) = \sum_{\forall v_i \in A(PCD_{II})} x^{\left(2^{d_{v_i}}-1\right)} \tag{33}$$

$$= \sum_{v_1 \in A(G)} x^{\left(2^1-1\right)} + \sum_{v_2 \in A(G)} x^{\left(2^2-1\right)} + \sum_{v_3 \in A(G)} x^{\left(2^3-1\right)} + \sum_{v_4 \in A(G)} x^{\left(2^4-1\right)} \tag{34}$$

$$= 4.2^n x + 8x^3 + 4(2^n + 3)x^7 + x^{15}. \tag{35}$$

$$G_{\mathfrak{f}}(PCD_{II}, x) = \sum_{v_i \in A(PCD_{II})} x^{\left(2^{2^{d_{v_i}}}-1\right)} \tag{36}$$

$$= \sum_{v_1 \in A(PCD_{II})} x^{\left(2^{2^1}-1\right)} + \sum_{v_2 \in A(PCD_{II})} x^{\left(2^{2^2}-1\right)} + \sum_{v_3 \in A(PCD_{II})} x^{\left(2^{2^3}-1\right)} + \sum_{v_4 \in A(PCD_{II})} x^{\left(2^{2^4}-1\right)} \tag{37}$$

$$= 4.2^n x^3 + 8x^{15} + 4(2^n + 3)x^{255} + x^{65536}. \tag{38}$$

Taking the derivative of each polynomial from (34) and (37) at $x = 1$ yields the Ghani Mersenne topological index and the Ghani Fermat topological index.

Type of atoms	$d_{v_1} = 1$	$d_{v_2} = 2$	$d_{v_3} = 3$	$d_{v_4} = 4$
Number of atoms	4.2^n	8	$4(2^n + 3)$	1

Table 9.
Types of atom in porphyrin-cored dendrimers-II.

$$T_i(PCD_{II}) = \begin{cases} 32.2^n + 123, & \text{whenever } T_i = G_m, \\ 4(283.2^n + 17179), & \text{whenever } T_i = G_f. \end{cases}$$

8. Porphyrin-cored dendrimer-III

The porphyrin-cored dendrimer-III, denoted by PCD_{III}, is a single dendrimer molecule development stage in generation G_n with n indicating its stages. PCD_{III} is made up of eight identical branches and one central core, see **Figure 10**. In the central core, the valency of 4 atoms is 2, the valency of 20 atoms is 3, the valency of 1 atom is 4, and the valency of 40 atoms is 5, see **Figure 10**.

8.1 Results and discussion

In each branch of PCD_{III}, the total number of atoms are $2^{n+1} - 1$, in which the valency of 2^n atoms is 1 and the valency of $2^n - 1$ atoms is 3. At the conclusion, we get **Table 10**.

$$G_m(PCD_{III}, x) = \sum_{\forall v_i \in A(PCD_{III})} x^{(2^{d_{v_i}} - 1)} \tag{39}$$

$$= \sum_{v_1 \in A(PCD_{III})} x^{(2^1-1)} + \sum_{v_2 \in A(PCD_{III})} x^{(2^2-1)} + \sum_{v_3 \in A(PCD_{III})} x^{(2^3-1)} + \sum_{v_4 \in A(PCD_{III})} x^{(2^4-1)} \tag{40}$$

$$= 8.2^n x + 4x^3 + 4(2.2^n + 3)x^7 + x^{15}. \tag{41}$$

$$G_f(PCD_{III}, x) = \sum_{v_i \in A(PCD_{III})} x^{\left(2^{2^{d_{v_i}}} - 1\right)} \tag{42}$$

$$= \sum_{v_1 \in A(PCD_{III})} x^{\left(2^{2^1}-1\right)} + \sum_{v_2 \in A(PCD_{III})} x^{\left(2^{2^2}-1\right)} + \sum_{v_3 \in A(PCD_{III})} x^{\left(2^{2^3}-1\right)} \tag{43}$$

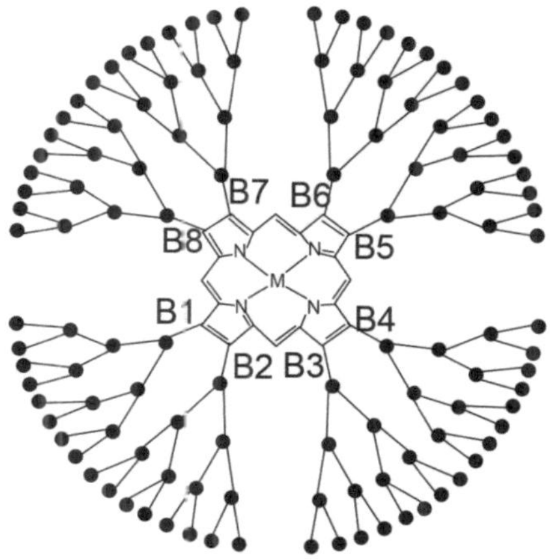

Figure 10.
Porphyrin dendrimer.

Type of atoms	$d_{v_1}=1$	$d_{v_2}=2$	$d_{v_3}=3$	$d_{v_4}=4$
Number of atoms	8.2^n	4	$4(2.2^n+3)$	1

Table 10.
Types of atoms in porphyrin-cored dendrimers-III.

$$= \sum_{v_4 \in A(PCD_{III})} x^{\left(2^{2^4}-1\right)} \quad (44)$$

$$= 8.2^n x^3 + 4x^{15} + 4(2.2^n+3))x^{255} + x^{65536}. \quad (45)$$

Taking the derivative of each polynomial from (40) and (43) at $x=1$ yields the Ghani Mersenne topological index and the Ghani Fermat topological index.

$$T_i(PCD_{III}) = \begin{cases} 4(16.2^n+12), & \text{whenever } T_i = G_{\mathrm{m}}, \\ 16(129.2^n+4291), & \text{whenever } T_i = G_{\mathfrak{f}}. \end{cases}$$

9. Porphyrin-cored dendrimer-IV

Our final dendrimer molecule (PCD_{IV}) is a porphyrin-cored dendrimer-IV (PCD_IV) with a $n \geq 1$ growth stage of generation $G_n.PCD_{IV}$ is composed of twelve identical branches and one central core, as illustrated in **Figure 11**. In the central core, which has the valency of 24 atoms is 3, the valency of 1 atom is 4, and 44 atom-bonds, see **Figure 11**.

9.1 Results and discussion

In each branch of PCD_{IV}, the total number of atoms are $2^{n+1}-1$, the valency of 2^n atoms is 1, and the valency of 2^n-1 atoms is 3. At the conclusion, we get **Table 11**.

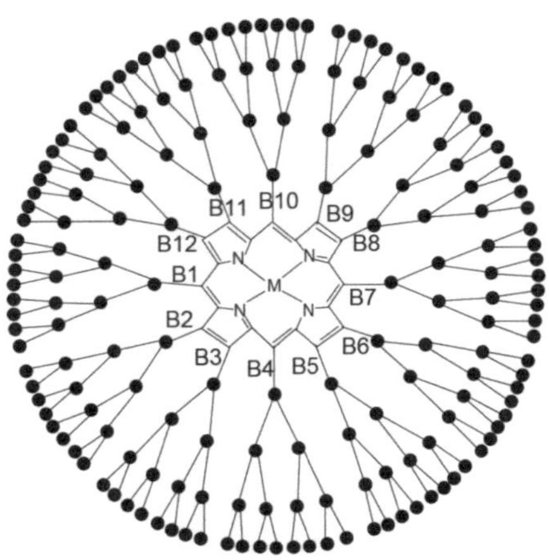

Figure 11.
Porphyrin dendrimer.

Type of atoms	$d_{v_2} = 1$	$d_{v_2} = 3$	$d_{v_3} = 4$
Number of atoms	3.2^{n+2}	$3.2^{n+2} + 12$	1

Table 11.
Types of atoms in porphyrin-cored dendrimers-IV.

$$G_m(PCD_{IV}, x) = \sum_{\forall v_i \in A(PCD_{IV})} x^{\left(2^{d_{v_i}} - 1\right)} \tag{46}$$

$$= \sum_{v_1 \in A(PCD_{IV})} x^{\left(2^1 - 1\right)} + \sum_{v_2 \in A(PCD_{IV})} x^{\left(2^3 - 1\right)} + \sum_{v_3 \in A(PCD_{IV})} x^{\left(2^4 - 1\right)} \tag{47}$$

$$= 3.2^{n+2}x + \left(3.2^{n+2} + 12\right)x^7 + x^{15}. \tag{48}$$

$$G_f(PCD_{IV}, x) = \sum_{v_i \in A(PCD_{IV})} x^{\left(2^{2^{d_{v_i}}} - 1\right)} \tag{49}$$

$$= \sum_{v_1 \in A(PCD_{IV})} x^{\left(2^{2^1} - 1\right)} + \sum_{v_2 \in A(PCD_{IV})} x^{\left(2^{2^3} - 1\right)} + \sum_{v_3 \in A(PCD_{IV})} x^{\left(2^{2^4} - 1\right)} \tag{50}$$

$$= 3.2^{n+2}x^3 + \left(3.2^{n+2} + 12\right)x^{255} + x^{65536}. \tag{51}$$

Taking the derivative of each polynomial from (47) and (50) at $x = 1$ yields the Ghani Mersenne topological index and the Ghani Fermat topological index.

$$T_i(PCD_{IV}) = \begin{cases} 3\left(8.2^{n+2} + 33\right), & \text{whenever } T_i = G_m, \\ 2\left(387.2^{n+2} + 34298\right), & \text{whenever } T_i = G_f. \end{cases}$$

10. Conclusion

This chapter focuses on eight important molecules: (i) poly or propyl ether-imine dendrimer, (ii) polyethylene amido amine dendrimer, (iii) porphyrin dendrimer, (iv) zinc porphyrin dendrimer, (v) porphyrin-cored dendrimer-I, (vi) porphyrin-cored dendrimer-II, and (vii) porphyrin-cored dendrimer-III, porphyrin-cored dendrimer-IV. Taking the first derivative at $x = 1$ from these polynomials yields the Ghani Mersenne index and the Ghani Fermat index. We compute the accurate formulae of the Ghani Mersenne index and Ghani Fermat index by using the technique of atom-bond partitioning of these eight molecules. Our investigated results, such as the G_m-index and G_f-index, always belong to a set of positive integers, not in decimal form, which is why they are useful for determining the physio-chemical properties of all molecules used in chemistry. Our findings also contribute significantly to network science by laying the groundwork for understanding the deep topology of (i) poly or propyl ether-imine dendrimer, (ii) polyethylene amido amine dendrimer, (iii) porphyrin dendrimer, (iv) zinc porphyrin dendrimer, and (v) porphyrin-cored dendrimer-I, II, III, and IV. These discoveries might help researchers discover how these molecules react with other materials in electronics and manufacturing.

Acknowledgements

This research was supported by the researchers Supporting Project Number (RSP2023R401), King Saud University, Riyadh, Saudi Arabia.

Conflict of interest

The authors declare no conflict of interest.

Author details

Muhammad Usman Ghani[1†], Syed Ajaz K. Kirmani[2†], Faisal Sultan[1*†] and Fairouz Tchier[3†]

1 Institute of Mathematics, Khawaja Fareed University of Engineering and Information Technology, Rahim Yar Khan, Pakistan

2 Department of Electrical Engineering, College of Engineering, Qassim University, Buraydah, Saudi Arabia

3 Mathematics Department, College of Science, King Saud University, Riyadh, Saudi Arabia

*Address all correspondence to: faisalsultan222@gmail.com

† These authors contributed equally.

IntechOpen

© 2024 The Author(s). Licensee IntechOpen. This chapter is distributed under the terms of the Creative Commons Attribution License (http://creativecommons.org/licenses/by/4.0), which permits unrestricted use, distribution, and reproduction in any medium, provided the original work is properly cited.

References

[1] Ghani MU et al. A paradigmatic approach to find the valency-based K-Banhatti and redefined Zagreb entropy for niobium oxide and a metal–organic framework. Molecules. 2022;**27**(20):6975

[2] Ghani MU et al. Entropy related to K-Banhatti indices via valency based on the presence of C6H6 in various molecules. Molecules. 2023;**28**(1):452

[3] Ghani MU et al. Characterizations of chemical networks entropies by K-banhatii topological indices. Symmetry. 2023;**15**(1):143

[4] Imran M et al. Computation of entropy measures for metal-organic frameworks. Molecules. 2023;**28**(12): 4726

[5] Chu Y-M et al. Computation of Zagreb polynomials and Zagreb indices for benzenoid triangular & hourglass system. Polycyclic Aromatic Compounds. 2022:1-10

[6] Gutman I, Furtula B, Katanic V. Randic index and information. AKCE International Journal of Graphs and Combinatorics. 2018;**15**(3):307-312

[7] Al-Ahmadi B, Saleh A, Al-Shammakh W. Downhill Zagreb topological indices and Mdn-polynomial of some chemical structures applied for the treatment of COVID-19 patients. Open Journal of Applied Sciences. 2021;**10**(04):395

[8] Deng H, Yang J, Xia F. A general modeling of some vertex-degree based topological indices in benzenoid systems and phenylenes. Computers & Mathematics with Applications. 2011;**61** (10):3017-3023

[9] Alam A et al. Degree-based entropy for a non-Kekulean benzenoid graph. Journal of Mathematics. 2022;**2022**

[10] Das KC. Atom-bond connectivity index of graphs. Discrete Applied Mathematics. 2010;**158**(11):1181-1188

[11] Farrukh F, Farooq R, Farahani MR. Calculating some topological indices of SiO_2 layer structure. Journal of Informatics and Mathematical Sciences. 2016;**8**(3):181-187

[12] Ghani MU et al. Valency-based indices for some succinct drugs by using M-polynomial. Symmetry. 2023;**15**(3):603

[13] Furtula B, Gutman I. A forgotten topological index. Journal of Mathematical Chemistry. 2015;**53**(4): 1184-1190

[14] Ghorbani M, Zangi S, Amraei N. New results on symmetric division deg index. Journal of Applied Mathematics and Computing. 2021;**65**(1):161-176

[15] Ghani MU et al. Hex-derived molecular descriptors via generalized valency-based entropies. IEEE Access. 2023;**11**:42052-42068

[16] Gutman I. Geometric approach to degree-based topological indices: Sombor indices. MATCH Communications in Mathematical and in Computer Chemistry. 2021;**86**(1):11-16

[17] Naylor T. Comprehensive Polymer Science. Vol. 643. Oxford: Pergamon; 1989

[18] Mekapati SB, Hansch C. Comparative QSAR studies on bibenzimidazoles and terbenzimidazoles inhibiting topoisomerase I. Bioorganic & Medicinal Chemistry. 2001;**9**(11):2885-2893

[19] Natarajan V et al. Effect of electron-phonon interaction and valence band

edge shift for carrier-type reversal in layered ZnS/rGO nanocomposites. Journal of Colloid and Interface Science. 2021;**586**:39-46

[20] Prabhu S et al. Computational analysis of some more rectangular tessellations of kekulenes and their molecular characterizations. Molecules. 2023;**28**(18):6625

[21] Vukicevic D, Furtula B. Topological index based on the ratios of geometrical and arithmetical means of end-vertex degrees of edges. Journal of Mathematical Chemistry. 2009;**46**(4):1369-1376

[22] Wiener H. Structural determination of paraffin boiling points. Journal of the American Chemical Society. 1947;**69**(1):17-20

[23] Zakharov AB, Tsarenko DK, Ivanov VV. Topological characteristics of iterated line graphs in the QSAR problem: A multigraph in the description of properties of unsaturated hydrocarbons. Structural Chemistry. 2021;**32**(4):1629-1639

[24] Zhong L. The harmonic index for graphs. Applied Mathematics Letters. 2012;**25**(3):561-566

[25] Arockiaraj M et al. QSPR analysis of distance-based structural indices for drug compounds in tuberculosis treatment. Heliyon. 2024;**10**(2)

[26] Alali AS et al. Algebraic structure graphs over the commutative ring Z m: Exploring topological indices and entropies using M-polynomials. Mathematics. 2023;**11**(18):3833

[27] Shpiz GB, Kryukov AP. The method of colored graphs for simplifying expressions with indices. Programming and Computer Software. 2021;**47**(1):25-28

[28] Zhou B, Gutman I. Further properties of Zagreb indices. MATCH Communications in Mathematical and in Computer Chemistry. 2005;**54**(1):233-239

[29] Xavier DA et al. Computing molecular descriptors of boron icosahedral sheet. International Journal of Quantum Chemistry. 2024;**124**(13):e27443

[30] Zuo X et al. Computing topological indices for molecules structure of polyphenylene via M-polynomials. Polycyclic Aromatic Compounds. 2022;**42**(4):1103-1112

[31] Rosary MS. On reverse valency based topological indices of metal–organic framework. Polycyclic Aromatic Compounds. 2021:1-14

[32] Zhou H-C, Long JR, Yaghi OM. Introduction to metal–organic frameworks. Chemical Reviews. 2012;**112**(2):673-674

[33] Li M et al. Topological analysis of metal–organic frameworks with polytopic linkers and/or multiple building units and the minimal transitivity principle. Chemical Reviews. 2014;**114**(2):1343-1370

Chapter 7

Slepian Pollak Inequality for Multi-Band Wavelets

Mohammad Younus Younus Bhat, Altaf Ahmad Bhat and Javaid Gani Dar

Abstract

The aim of this article is to study Slepian Pollak Inequality (Bell Labs inequality) for multi-band wavelets (M-band wavelets). A version of the already known energy inequality for the wavelets helps to zoom in onto narrow-band high-frequency components of a signal.

Keywords: wavelet, M-band, Slepian Pollak Inequality, Fourier transform, wavelet frame

1. Introduction

To study the high-frequency signals whose bandwidth is low, the classical orthogonal wavelets are not enough. As a result of this, M-band novel orthonormal wavelets were born [1, 2]. The reason behind the greater factor $M(M > 2)$ is that, instead of the standard decomposition which gives logarithmic frequency resolution, the new wavelet decomposition gives rise to a combination of logarithmic and linear frequency resolution and thus designs a more compatible tiling of the time-frequency plane which is absent in 2-band wavelets. For more difference between the novel multi-band wavelets and classical 2-band wavelets, we refer to [3] and the references therein.

Moving to other side, tight wavelet frames have redundancy which is absent in the orthonormal wavelets. By ignoring orthonormality and keeping redundancy, it becomes easier to construct tight wavelet frames that the classical orthonormal wavelets. This construction becomes easier when we use the unitary extension principle (UEP) as catalyst [4]. The research in the area of frame theory is well developed. To mention a few, we refer to [5–7].

The fast growing and emerging branch of research wavelet analysis has became a bridge to connect mathematicians with other researchers like physicians, chemists, engineers, etc. Hence, it becomes quite interesting to study different aspects like uncertainty principle in the context of wavelet analysis. The uncertainty principle inequalities are the basic blocks of signal analysis. Its classical version was studied by Heisenberg, Pauli, Weyl, and Wiener. Going to the application side it is here worth to mention the Slepian Pollak energy inequality (or Bell Labs inequality) and the D. Gabor on a windowed transform. For more details about these inequalities we refer to [8–11].

 IntechOpen

From the motivational point of view, we consider P as projection operator on wavelet which is in fact subspace of $L^2(\mathbb{R})$. For any signal $f \in L^2(\mathbb{R})$ with finite energy, we consider a time-truncation operator T, which in mathematical terms is defined by $Tf(t) = f(t)\chi_{[a,b]}(t)$, with usual characteristic function $\chi_{[a,b]}$ on any segment $[a, b]$. Let the signal have unit norm, i.e. $\|f\|_2 = 1$ and $\|Tf\|_2$ has certain value in $[0, 1]$. Then the authors in [11] considered the possible values of $\|Pf\|_2$, which is in fact a version of uncertainty principle as there is uncertainty between time-truncation operator T and projection operator P. To be precise they solved Fourier projection transform and expressed it in terms of well-known Bell Labs inequality.

Motivated by the work of [11], we consider its generalisation on M-band wavelets. The rest of paper is organised as follows. The paper is organised in the following manner. In Section 2, we take some review of multi-band wavelets. In Section 3, we establish the Slepian Pollak energy inequality.

2. Preliminaries on multi-band wavelet frames

Let us begin with some preliminaries on M-band wavelet frames. We make use of the symbols $\mathbb{N}, \mathbb{N}_0, \mathbb{Z}$, and $\mathbb{R}$ for the sets of all natural numbers, nonnegative integers, integers, and real numbers, respectively. Whereas the identity matrix of size $M > 2$ is denoted by I_M.

For any function $f \in L^1(\mathbb{R})$, we define its Fourier transform as:

$$\hat{f}(\zeta) = \int_{\mathbb{R}} f(x) e^{-i\zeta x} dx, \quad \zeta \in \mathbb{R}$$

and its inverse is

$$f(x) = \frac{1}{2\pi} \int_{\mathbb{R}} \hat{f}(\zeta) e^{i\zeta x} d\zeta, \quad x \in \mathbb{R}.$$

For given $\mathfrak{G} := \{\mathfrak{g}_1, \dots, \mathfrak{g}_L\} \subset L^2(\mathbb{R})$, define the M-band wavelet system

$$X(\mathfrak{G}) := \left\{ \mathfrak{g}^{\ell}_{j,k} : 1 \le \ell \le L; j, k \in \mathbb{Z} \right\} \tag{1}$$

where $\mathfrak{g}^{\ell}_{j,k} = M^{j/2} \mathfrak{g}^{\ell}\left(M^j . - k\right)$. The wavelet system $X(\mathfrak{G})$ is treated as the *M-band wavelet frame*, or simply a *wavelet frame*, whenever we have positive numbers $0 < \mathfrak{A} \le \mathfrak{B} < \infty$ such that

$$\mathfrak{A}\|f\|_2^2 \le \sum_{\ell=1}^{L} \sum_{j \in \mathbb{Z}} \sum_{k \in \mathbb{Z}} \left| \left\langle f, \mathfrak{g}^{\ell}_{j,k} \right\rangle \right|^2 \le \mathfrak{B}\|f\|_2^2. \tag{2}$$

holds for all $f \in L^2(\mathbb{R})$.

The constant $\mathfrak{A}$ which is in is called fact largest the *lower* frame bound, whereas the smallest constant $\mathfrak{B}$ is treated as the *upper wavelet frame bounds*. We have a *tight wavelet frame* whenever $\mathfrak{A} = \mathfrak{B}$. Moreover, the generated system will be a *Parseval wavelet frame* if $\mathfrak{A} = \mathfrak{B} = 1$. Here, every function $f \in L^2(\mathbb{R})$ is reshaped as

$$f(x) = \sum_{\ell=1}^{L} \sum_{j \in \mathbb{Z}} \sum_{k \in \mathbb{Z}} \left\langle f, \mathfrak{g}_{j,k}^{\ell} \right\rangle \mathfrak{g}_{j,k}^{\ell}(x). \tag{3}$$

Framelet systems are often constructed from the construction of MRA, which is in turn depends on refinable functions. We say function $\mathfrak{f} \in L^2(\mathbb{R})$ an M-refinable whenever it satisfies a refinement equation:

$$\mathfrak{f}(x) = \sum_{k \in \mathbb{Z}} h_0[k] \mathfrak{f}(Mx - k). \tag{4}$$

for some $h_0 \in l^2(\mathbb{Z})$. Its Fourier transform is

$$\hat{\mathfrak{f}}(\zeta) = m_0\left(\frac{\zeta}{M}\right) \hat{\mathfrak{f}}\left(\frac{\zeta}{M}\right), \tag{5}$$

where

$$m_0(\zeta) = \frac{1}{M} \sum_{k \in \mathbb{Z}} h_0[k] e^{ik\zeta},$$

is a 2π-periodic measurable function in $L^\infty[-\pi, \pi]$ and is often called the *refinement symbol* of $\mathfrak{f}$. We further assume that:

$$\lim_{\zeta \to 0} \hat{\mathfrak{f}}(\zeta) = 1, \quad \text{and} \quad \sum_{k \in \mathbb{Z}} \left| \hat{\mathfrak{f}}(\zeta + 2k\pi) \right|^2 \in L^\infty[-\pi, \pi]. \tag{6}$$

With $\mathfrak{f} \in L^2(\mathbb{R})$ a refinable and compactly supported function, consider the space V_0, which is created by $\{\mathfrak{f}(\cdot - k) : k \in \mathbb{Z}\}$ and $V_j = \{\mathfrak{f}(M^j \cdot) : \mathfrak{f} \in V_0\}, j \in \mathbb{Z}$. Obviously if $\mathfrak{f}$ is compactly supported, the $\{V_j\}_{j \in \mathbb{Z}}$ ends with an MRA. Note that MRA is a collection of closed subspaces $\{V_j\}_{j \in \mathbb{Z}}$ of $L^2(\mathbb{R})$ which obeys: (i) $V_j \subset V_{j+1}, j \in \mathbb{Z}$, (ii) $\cup_{j \in \mathbb{Z}} V_j$ is dense in $L^2(\mathbb{R})$, and (iii) $\cap_{j \in \mathbb{Z}} V_j = \{0\}$. Let $\mathfrak{G} := \{\mathfrak{g}_1, \ldots, \mathfrak{g}_L\} \subset V_1$, then

$$\hat{\mathfrak{g}}_\ell(\zeta) = m_\ell\left(\frac{\zeta}{M}\right) \hat{\mathfrak{f}}\left(\frac{\zeta}{M}\right), \tag{7}$$

where

$$m_\ell(\zeta) = \frac{1}{M} \sum_{k \in \mathbb{Z}} h_\ell[k] e^{ik\zeta}, \quad \ell = 1, \ldots, L$$

are the 2π-periodic measurable functions in $L^\infty[-\pi, \pi]$ which are named as *framelet symbols*. The well-known catalyst *unitary extension principle* (UEP) incorporates a sufficient condition on the system $\mathfrak{G}$ so that the resulting M-band system $X(\mathfrak{G})$ becomes a tight frame of $L^2(\mathbb{R})$. It were Han and Cheng [6] who provided a complete characterisation of these multi-band tight wavelet frames using the machinery of the unitary extension principle. In fact they used the following tool.

Theorem 2.1. [6] Suppose that the refinable function $\mathfrak{f}$ and the framelet symbols $m_0, m_1, \dots, m_L$ satisfy (4)–(7). Define $\mathfrak{g}_1, \dots, \mathfrak{g}_L$ by (7). Let $\mathcal{M}(\zeta) = \left\{m_\ell\left(\zeta + \frac{2\pi p}{M}\right)\right\}_{\ell,p=0}^{M-1}$ such that $\mathcal{M}(\zeta)\mathcal{M}^*(\zeta) = I_M$, for $a.e$ $\zeta \in \sigma(V_0) := \left\{\zeta \in [-\pi, \pi] : \sum_{k\in\mathbb{Z}} \left|\hat{\mathfrak{f}}(\zeta + 2k\pi)\right|^2 \neq 0\right\}$, then M-band wavelet system $X(\mathfrak{g})$ forms a tight wavelet frame for $L^2(\mathbb{R})$ with frame bound 1.

3. Projection on MRA spaces

Define the projection $P : f \to f_m \in V_m$ by

$$Pf(x) = f_m(x) = \sum_{\ell=1}^{L}\sum_{j=-\infty}^{m-1}\sum_{k=-\infty}^{\infty}\left\langle f, \mathfrak{g}_{j,k}^{\ell}\right\rangle \mathfrak{g}_{j,k}^{\ell}(x) = \sum_{k=-\infty}^{\infty}\left\langle f, \mathfrak{f}_{j,k}\right\rangle \mathfrak{f}_{j,k}(x). \tag{8}$$

Lemma 3.1. *Let $\mathfrak{f}$ be a continuous scaling function of bounded variation on $\mathbb{R}$. Suppose one of the following equations hold*

$$\sum_{k\in\mathbb{Z}}\left|\hat{\mathfrak{f}}(\zeta + 2\pi k)\right| < A < \infty, \quad \text{for a.e.}\zeta \in \mathbb{R}, \tag{9}$$

$$\sum_{k\in\mathbb{Z}}\left|\mathfrak{f}(x - k)\right| < B < \infty, \quad \text{for a.e.}x \in \mathbb{R}. \tag{10}$$

Then for any integer j and $f \in L^2(\mathbb{R})$, we have

$$f_j(x) = \frac{1}{2\pi}\int_{-\infty}^{\infty} \hat{f}(\zeta)\hat{\mathfrak{f}}(M^{-j}\zeta)e^{i\zeta x}\sum_{n=-\infty}^{\infty}\hat{\mathfrak{f}}(M^{-j}\zeta + 2\pi n)e^{i2\pi n M^j x}d\zeta. \tag{11}$$

Proof. Using Parseval's identity, we have from Eq. (8)

$$\begin{aligned}
f_j(x) &= \frac{1}{2\pi}\sum_{k=-\infty}^{\infty}\int_{-\infty}^{\infty}\hat{f}(\zeta)\overline{\hat{\mathfrak{f}}(M^{-j}\zeta)}e^{ikM^{-j}\zeta}\mathfrak{f}(M^j x - k)d\zeta \\
&= \frac{1}{2\pi}\int_{-\infty}^{\infty}\hat{f}(\zeta)\overline{\hat{\mathfrak{f}}(M^{-j}\zeta)}\sum_{k=-\infty}^{\infty}\left\{e^{ikM^{-j}\zeta}\mathfrak{f}(M^j x - k)\right\}d\zeta \\
&= \frac{1}{2\pi}\int_{-\infty}^{\infty}\hat{f}(\zeta)\overline{\hat{\mathfrak{f}}(M^{-j}\zeta)}e^{i\zeta x}\sum_{k=-\infty}^{\infty}\left\{e^{-i(M^j x - k)M^{-j}\zeta}\mathfrak{f}(M^j x - k)\right\}d\zeta \\
&= \frac{1}{2\pi}\int_{-\infty}^{\infty}\hat{f}(\zeta)\overline{\hat{\mathfrak{f}}(M^{-j}\zeta)}e^{i\zeta x}\sum_{n=-\infty}^{\infty}\left\{e^{i2\pi n M^j x}\hat{\mathfrak{f}}(M^{-j}\zeta + 2\pi n)\right\}d\zeta,
\end{aligned}$$

where we have used Poisson Summation formula, because of Eq. (10). This completes the proof. □

Lemma 3.2. *Let $\mathfrak{f}$ be as in Lemma 3.1 and P be the projection operator. If $\mathcal{F}$ is a bounded family of square integrable functions, then its image $P(\mathcal{F})$ is uniformly bounded family on $\mathbb{R}$.*

Proof. Let $\mathfrak{f}$ satisfy condition (9). Then from (11), we have by Schawrtz Inequality

$$
\begin{aligned}
|f_j(x)| &\leq \frac{1}{2\pi}\int_{-\infty}^{\infty} |\hat{f}(\zeta)||\hat{\mathfrak{f}}(M^{-j}\zeta)|\left|\sum_{n=-\infty}^{\infty} \hat{\mathfrak{f}}(M^{-j}\zeta+2\pi n)e^{i(\zeta+2\pi nM^j)x}\right|d\zeta \\
&\leq \frac{1}{2\pi}\|\hat{f}\|_2\left\{\int_{-\infty}^{\infty}|\hat{\mathfrak{f}}(M^{-j}\zeta)|^2\left(\sum_{n=-\infty}^{\infty}|\hat{\mathfrak{f}}(M^{-j}\zeta+2\pi n)|\right)^2 d\zeta\right\}^{1/2} \\
&\leq \frac{1}{\sqrt{2\pi}}A\|f\|_2\left(\int_{-\infty}^{\infty}|\hat{\mathfrak{f}}(M^{-j}\zeta)|^2 d\zeta\right)^{1/2} \\
&\leq M^{j/2}A\|f\|_2.
\end{aligned}
$$

Moreover, if $\mathfrak{f}$ satisfies Eq. (10), then we have

$$
|f_j(x)| \leq \frac{1}{2\pi}\int_{-\infty}^{\infty} |\hat{f}(\zeta)||\hat{\mathfrak{f}}(M^{-j}\zeta)|\left|\sum_{n=-\infty}^{\infty} \mathfrak{f}(M^j x-k)\right|d\zeta \leq M^{j/2}B\|f\|_2.
$$

Theorem 3.3. *Let* $\mathfrak{f}$ *satisfy the conditions of Lemma 3.1, P be the projection operator, and T be the truncation operator. Then PT is a compact operator.*

Proof. Define

$$
\Delta_f(\epsilon) = \sup_x |f(x+\epsilon) - f(x)|.
$$

where $\epsilon > 0$. Let $\mathfrak{f}$ satisfy the condition (9), then

$$
\begin{aligned}
\Delta_{f_j}(\delta) &\leq \int_{-\infty}^{\infty}|\hat{f}(\zeta)||\hat{\mathfrak{f}}(M^{-j}\zeta)|\sum_{n=-\infty}^{\infty}|\hat{\mathfrak{f}}(M^{-j}\zeta+2\pi n)|\left|e^{i(\zeta+2\pi nM^j)\epsilon}-1\right|d\zeta \\
&\leq \frac{\|f\|_2}{\sqrt{2\pi}}\left\{\int_{-\infty}^{\infty}\left(|\hat{\mathfrak{f}}(M^{-j}\zeta)|\sum_{n=-\infty}^{\infty}|\hat{\mathfrak{f}}(M^{-j}\zeta+2\pi n)|\left|e^{i(\zeta+2\pi nM^j)\epsilon}-1\right|\right)^2 d\zeta\right\}^{1/2}.
\end{aligned}
$$

Since

$$
|\hat{\mathfrak{f}}(M^{-j}\zeta)|^2\left(\sum_{n=-\infty}^{\infty}|\hat{\mathfrak{f}}(M^{-j}\zeta+2\pi n)|\left|e^{i(\zeta+2\pi nM^j)\epsilon}-1\right|\right)^2 \leq 4A^2|\hat{\mathfrak{f}}(M^{-j}\zeta)|^2 \in L^1(\mathbb{R})
$$

and by (9), the series

$$
\sum_{n=-\infty}^{\infty}|\hat{\mathfrak{f}}(M^{-j}\zeta+2\pi n)|\left|e^{i(\zeta+2\pi nM^j)\delta}-1\right| \to 0 \quad \text{as} \quad \epsilon \to 0,
$$

Lebesque Dominated convergence theorem implies that

$$
\lim_{\epsilon\to 0}\Delta_{f_j}(\epsilon) = 0.
$$

By taking inverse Fourier transform to the above, the result also holds. Thus for any bounded family $\mathcal{F}$ in $L^2(\mathbb{R})$, its image $P(\mathcal{F})$ is equicontinuous and uniformly bounded family on $\mathbb{R}$. Hence, according to Ascoli's Theorem, PT is a compact operator.

Thus with the above results in hand we are in a position to say that we have proved the $M-$band version of Bell Labs inequality. To be precise, mathematically we have proved the following result:

Let $\mathfrak{f}$ be as in the Lemma 2, P be the projection operator, and T be the truncation operator, then (P, T) is a pair satisfying Bell Labs inequality

$$\arccos\frac{\|Tf\|_2}{\|f\|_2} + \arccos\frac{\|Pf\|_2}{\|f\|_2} \geq \arccos\|Tf\|_2.$$

Acknowledgements

We are deeply indebted to the referee for his/her valuable suggestions which greatly improved the presentation of this paper.

Conflict of interest

The authors declare no conflict of interest.

Data availability

Data will be available on the request to authors.

Author details

Mohammad Younus Younus Bhat[1*], Altaf Ahmad Bhat[2] and Javaid Gani Dar[3]

1 Department of Mathematical Sciences, Islamic University of Science and Technology, Kashmir, India

2 Sciences and Mathematics Unit, University of Technology and Applied Sciences, Salalah, Sultanate of Oman

3 Department of Applied Sciences, Symbiosis Institute of Technology, Symbiosis International (Deemed University), Pune, India

*Address all correspondence to: gyounusg@gmail.com

IntechOpen

© 2025 The Author(s). Licensee IntechOpen. This chapter is distributed under the terms of the Creative Commons Attribution License (http://creativecommons.org/licenses/by/4.0), which permits unrestricted use, distribution, and reproduction in any medium, provided the original work is properly cited.

References

[1] Shah FA, Bhat MY. On framelet kernels of M-band wavelet frames. Gulf Journal of Mathematics. 2015;3(4):59-66

[2] Steffen P, Heller PN, Gopinath RA, Burrus CS. Theory of regular M-band wavelet bases. IEEE Transactions on Signal Processing. 1993;**41**(12): 3497-3510

[3] Bi N, Dai X, Sun Q. Construction of compactly supported M-band wavelets. Applied and Computational Harmonic Analysis. 1999;**6**:113-131

[4] Ron A, Shen Z. Affine systems in $L^2(R^d)$: The analysis of the analysis operator. Journal of Functional Analysis. 1997;**148**:408-447

[5] Chai A, Shen Z. Deconvolution: A wavelet frame approach. Numerische Mathematik. 2007;**106**:529-587

[6] Han JC, Cheng Z. The construction of M-band tight wavelet frames. In: Proceedings of the Third Intemational Conference on Machine Learning and Cybemetics; Shanghai. 2004. pp. 26-29

[7] Rakotomamonjy A, Canu S. Frames, reproducing kernels, regularization and learning. Journal of Machine Learning Research. 2005;**6**:1485-1515

[8] Benedetto JJ. Uncertainty principle inequalities and spectrum estimation. In: Byrnes JS, Byrnes JL, editors. Recent Advances in Fourier Analysis and Its Applications. Dordrecht: Iiluwer Acad. Publ; 1990. pp. 143-182

[9] Folland GB, Sitaram A. The uncertainty principles: A mathematical survey. Journal of Fourier Analysis and Applications. 1997;**106**:209-238

[10] Havin V, Joricke B. The Uncertainty Principle in Harmonic Analysis, A Series of Modern Survey in Mathematics No 28. Berlin: Springer-Verlag; 1994

[11] Karanikas K, Proios G. The slepian-pollak energy inequality for wavelets and frames. Applicable Analysis. 1991;**70**: 321-329

Section 3

Perspectives on Proof, Education, and Pedagogy

Chapter 8

Perspective Chapter: Proofs in Number Theory – Audiences, Purposes, and Modes

Tim Rowland

Abstract

David Tall identified three 'worlds of mathematics', and in this chapter, we consider, with examples, the nature of proof in number theory in each of those worlds. While the role of conventional, formal proofs in establishing the canon of 'what we know' are well-understood, the approach to proof *via* generic example is worthy of serious consideration. Within mathematics, number theory offers considerable potential for the presentation of a general argument through the medium of a particular example. I demonstrate this by reference to an episode in an undergraduate mathematics lecture, in which the mode of instruction offers insight into the genetic origins of the topic under consideration.

Keywords: proof, conjecture, teaching, explanation, generic examples

1. Introduction

In this chapter, I advocate the use of particular-but-generic proof strategies in undergraduate classrooms and in textbooks, in order to convince students of the truth of number-theoretic theorems and student-generated conjectures. The domain of number theory lends itself particularly well to generic argument, presented with the intention of conveying the force and the structure of a conventional generalised argument through the medium of a particular case. The potential of the generic example as a didactic tool is virtually unrecognised, although the use of such examples has good historical provenance.

2. Audiences, proofs

The desire to perceive regularities, general trends that extend beyond particular instances, is a natural human propensity. It is useful to know the 'shape' of a typical day during the week, and essential to know what will happen as I move my feet between pedals when I'm driving a car. Mathematics is distinguished by the complexity of such regularities, and the insight necessary to articulate them in conjectures; and also, by our grounds for confidence in the 'truth' of such conjectures. In most domains of

IntechOpen

human knowledge, those grounds will be confidence in the consistency of empirical data ('the car always slows down when I push one particular foot-pedal'). This might be strengthened by an explanation of some kind (the design of the braking system), or a fundamental explanatory theory. Such theories hold good for as long as what we observe is consistent with their predictions.

The notion of 'proof' is particular to mathematics. Proof proceeds from known truths (some of them called 'theorems') to confirm the general validity of conjectures, by a process of logical argument. For more than two millennia, Euclid's *Elements* has offered the paradigm example of the role of proof in mathematics. Even within experimental science, the primary criterion and standard of evaluation of scientific theory is evidence, not proof.

In the early years of this millennium, David Tall considered the different ways in which statements are validated in what he called 'the three worlds of mathematics' (e.g. [1]). Tall's approach builds on Jerome Bruner's three modes of representation—enactive, iconic and symbolic [2]. Tall names his three worlds embodied, symbolic and formal. Briefly, the embodied world is the fundamental human mode of operation based on perception and action. The symbolic-proceptual world is a world of mathematical symbol-processing, and the formal-axiomatic world involves a shift into formalism ([1], p. 3). I shall consider the first and the last of these three worlds, drawing on two episodes from my own mathematical and pedagogical activity in the last three decades.

2.1 Formal proof

In December 1995, I spent a week as a hospital inpatient, following straightforward and successful surgery. Various visitors came to see me: to my surprise, one of them was one particular Cambridge colleague. It would be fair to say that Bob Hall and I shared very few interests beyond mathematics. In an attempt to 'make conversation' with Bob, I spoke about some Number Theory questions that I'd been setting for the examinations to be taken the following May. One was on primitive Pythagorean Triples (x, y, z) where x, y, z have no common non-unit factor, drawing on the 'classical form' $(2mn, m^2\text{-}n^2, m^2 + n^2)$, where m, n are of opposite parity, and coprime. Now, Euclid 'knew' and (some 500 years later) Diophantus proved that every primitive Pythagorean triple is classical. But, as our conversation progressed, we found that, when we taught even the more-advanced undergraduate course, both of us encouraged our students to engage in a process of active enquiry: to find examples and to generate conjectures about the form of primitive Pythagorean triples. But whilst it is certainly possible, given sufficient time and appropriate prompting, for students to detect and formulate a classical form for all primitive Pythagorean triples in this way, the standard proof that all primitive Pythagorean triples are classical (e.g. [3]) takes a different approach. As Bob and I talked, we developed a possible approach to a proof that was a more natural development of the earlier enquiry. After an hour or so, Bob left, but a week or so later we had agreed on a text that faithfully captured our approach, within a 'formal' proof. That proof began with a definition which facilitated the statement and presentation of the subsequent argument, as follows:

Definition: Let α, β be two Pythagorean triples (x, y, z), (x', y', z') respectively. α will be said to *precede* β (notation $\alpha \leq \beta$) if there is some positive integer r such that $x'/x = y'/y = z'/z = r$. In this case, write $\beta = r\alpha$.

For this relation to hold, it is clearly necessary, but not sufficient, that the corresponding right-angled triangles are similar. Thus, $\leq$ is a partial ordering on the set of Pythagorean triples in which the minimal elements (those with no predecessors in this partial ordering) are clearly primitive Pythagorean triples.

DOI: http://dx.doi.org/10.5772/intechopen.1012302

Lemma: Every predecessor of a classical Pythagorean triple is classical (i.e. the property 'classical' is downward closed).

The Lemma is then shown to enable the proof that every primitive Pythagorean Triple is classical. It was published in Ref. [4], sitting comfortably in Tall's world of formal mathematics, proceeding from 'agreed' starting points to the 'new' result, *via*—albeit implicitly—the conventions of first-order predicate logic with quantifiers, including rules of inference such as $(x)(A(x) \rightarrow A(n)$. The unspoken agreement is that a suitably-equipped machine, such as the Rocq Prover (https://rocq-prover.org/) could confirm, or otherwise, the validity of the argument.

3. Procedures and proofs

The use of examples to point to abstract concepts and to general procedures is commonplace pedagogical practice, both within and beyond mathematics [5, 6]. In the field of number theory, a case in point might be explication of the Euclidean algorithm for the greatest common divisor of two natural numbers. Beginning with, say, 194 and 40 the demonstration proceeds:

$$194 = 4 \times 40 + 34.$$

$$40 = 1 \times 34 + 6.$$

$$34 = 5 \times 6 + 4.$$

$$6 = 1 \times 4 + 2.$$

$$4 = 2 \times 2 + 0.$$

In order to apply the procedure to another pair of natural numbers, the student needs to become aware of the *status* of each number in each row of the procedure, and how each row *relates* to the next. That is, not only to agree that each line is a true statement, but to appreciate how it has been initiated and structured. They need, so to speak, to see the 'big picture' within the process being exemplified. As teacher, I might assist this by (say) underlining the quotients 4, 1, 5, 1, 2, in red. I might draw diagonal lines joining the divisor and remainder in each line to the dividend and divisor, respectively, in the next, for example joining the two 40s, the two 34 s, and so on. (It is relevant to pause to reflect on how you made sense of the previous sentence: perhaps the example was more illuminating than the description of the general procedure that followed it.) The choice of example (194, 40) was made in recognition of its merits in its own right and relative to some alternatives. I judge it to be preferable to (194, 48), which is a poor paradigm because, for that pair, the algorithm terminates too soon.

3.1 Embodied proof in generic examples

Much less common is the use of examples to explain why general relationships might hold: in short, to *prove*. One reason why this might be the case is clear enough- because one or more examples cannot prove a statement about an infinite category of cases. Yet there is a sense in which the *presentation* of a single example can speak for some general truth, and for some general argument above and beyond the particularities of the example itself. Such examples, suitably structured to be not just a confirming instance but a chain of reasoning, are known as generic examples. Mason and Pimm ([7], p. 284) explain that 'The generic proof, although given in terms of a particular number, nowhere relies on any specific properties of that number'. As Balacheff ([8], p. 219) so clearly and elegantly puts it:

The generic example involves making explicit the reasons for the truth of an assertion by means of operations or transformations on an object that is not there in its own right, but as a characteristic representative of the class.

The generic example serves not only to present a confirming instance of a proposition—which it certainly is—but to provide insight as to *why* the proposition holds true for that single instance. The transparent presentation of the example is such that analogy with other instances is readily achieved, and their truth is thereby made manifest. Ultimately the audience can conceive of no possible instance in which the analogy could not be achieved.

In considering and discussing the nature, and the role, of generic examples, the context shifts from the collaborative activity of mathematicians to an instructor (such as a teacher, a lecturer) attempting to enable others to 'see' a particular mathematical generalisation. In the sections which follow we look at the nature of generic examples, and instances of their application in number theory, in classroom settings.

4. Looking back

The story (probably apocryphal, but see ([9], pp. 60–62) for one version) is told about the child C. F. Gauss, who astounded his village schoolmaster by his rapid calculation of the sum of the integers from 1 to 100. Whilst the other pupils performed laborious column addition, Gauss added 1–100, 2–99, 3–98 and so on, and finally computed fifty 101 s with ease. The power of the story is that it offers the listener a means to add, say, the integers from 1 to 200. Gauss's method demonstrates, by generic example, that the sum of the first $2k$ positive integers is $k(2k+1)$. Nobody who could follow Gauss' method in the case $k = 50$ could possibly doubt the general case. It is important to emphasise that it is not simply the *fact* that the proposition $1 + 2 + 3 + \ldots + 2k = k(2k+1)$ has been verified as true in the case $k = 50$. It is the *manner* in which it is verified, the form of presentation of the confirmation.

Paul Hoffman [10] recounts the story in his best-seller *The Man Who Loved Only Numbers*. His comment on it (quoting mathematician Ronald Graham) is a telling testimony to the genericity of Gauss' method.

What makes Gauss' method so special .... Is that it does not just work for this specific problem but can be generalised to find the sum of the first 50 integers or the first 1000 integers ... or whatever you want ([10], p. 208).

In introducing the notion 'generic example' to audiences of all kinds—undergraduate and graduate students, mathematics education conference-goers, 'general audiences'—I often choose Gauss' method as a paradigm of the genre. We should not be surprised that Gauss, of all people, provided it. Ironically, his *Disquisitiones Arithmeticae* established the 'modern' standard for generality in number theoretic proof arguments.

By contrast, Pierre de Fermat (1601–65) was notorious for stating number theoretic results in the absence of formal proof. In particular, it was Euler who gave a general proof of the 'Little' Theorem (if p is prime and a an integer, p divides $a^p - a$) some decades after Fermat stated it. Burn [11] offers some suggestions concerning the kinds of reasoning that Fermat himself might have used to establish the truth of some claims associated with his Little Theorem, made in a letter to Mersenne in 1640. These claims were developed in the course of Fermat's search for perfect numbers. A natural number (such as 6 or 28) is said to be *perfect* if it is equal to the sum of its proper divisors. Around 300 BC, Euclid had established that the set of perfect numbers can be identified with integers of the form $2^{n-1}(2^n - 1)$ where $2^n - 1$ is prime. In his letter to Mersenne (after whom such primes are named), Fermat claimed that if n is composite, then $2^n - 1$ is not prime. The proof amounts to the observation that $2^a - 1$ divides $2^{ab} - 1$. The converse, however, is false: in 1536, Hudalrichus Regius had shown that although 11 is prime, $2^{11}–1$ is composite, and so $2^{10}(2^{11}–1)$ is not a perfect number.

Fermat made a claim which was to transform the previously Herculean task of determining whether or not $2^p - 1$ is prime for a given prime p. In effect, Fermat claimed that if an integer of this form has a prime factor, then that factor is of the form $2kp + 1$ (the factor 1 is covered by $k = 0$, and it follows from Fermat's Little Theorem that $2^p - 1$ itself is of the form $2kp + 1$). Thus, to decide whether a proper factor of $2^{11}–1$ exists, we only need to consider 23, since this is the *only* prime of the form $22k + 1$ with square less than $2^{11}–1$. In fact, $2^{11}–1 = 23x89$. (Note that 89 is also of the form $22k + 1$, as expected).

In his letter, Fermat exemplifies this statement about prime factors of $2^p - 1$ with reference to this case, that is when $p = 11$. Burn [11] reconstructs the argument that Fermat might have given with reference to this particular-but-generic case. Burn then continues: 'Now we generalise the generic example of factorising $2^{11} - 1$ by expressing the argument algebraically'. Of course, this accords with good modern practice, although Burn does not suggest that Fermat, having established the generic, then required the general formulation to be convinced of the general case.

5. Teaching and learning Wilson's theorem

The obscure Cambridge mathematician John Wilson is remembered to this day on account of a theorem stated in 1770, a century after Fermat's demise:

p is prime if and only if $(p\text{-}1)! = -1 \pmod{p}$.

To be precise, it was Edward Waring, Isaac Newton's successor (and Stephen Hawking's predecessor) as the Cambridge Lucasian Professor of Mathematics, who stated the result of his former student, Wilson. In the best traditions of the time, neither Wilson nor Waring managed to prove the theorem: its status seems to have been a conjecture, the outcome of inductive reasoning from examples. It fell to Lagrange to give the first proof of Wilson's theorem, in 1773. How then, might we approach the genesis of the theorem and the construction of its proof with the hindsight and didactic insights of the twenty-first century? What might a generic proof of that theorem look like?

As a preliminary, we would need to know that ±1 are the only self-inverse elements under multiplication modulo p. Now consider the prime number 13 (17 or 19 would do equally well) and list the reduced set of residues modulo 13:

1 2 3 4 5 6 7 8 9 10 11 12

Pair each of the numbers from 2 to 11 with its (distinct) multiplicative inverse mod 13: (2, 7), (3, 9), (4, 10), (5, 8), (6, 11). 1 and 12, of course, are self-inverse. Clearly, the product of the integers from 2 to 11 must be congruent to 1^5, that is 1, modulo 13. Therefore $12! \equiv 1x1x12$ (=12) mod 13. The argument is generic, since 13 was in no way an untypical choice: the pairing would work equally well with any prime.

I now invite you to join me in a session [12] with a class of about 20 first-year undergraduate joint honours mathematics-with-education students. The course is an introduction to the Theory of Numbers, and the 'schedule' (syllabus) includes Wilson's theorem. I could have stated the theorem and proved it formally in 5 minutes. In fact, it took an hour to make some conjectures and to work on proof. This is what happened.

First, I asked them to evaluate 4! mod 5, 6! mod 7, 10! mod 11, and to write down a conjecture. The most common version of the conjecture was $n! \equiv n \bmod (n+1)$. The 'for all n' seemed to be implicit. I asked them to evaluate 5! mod 6. They did, and they were visibly surprised by the refutation. I asked whether they could modify the conjecture. At first, they homed in on the even/odd distinction between moduli, but $n = 8$ led to further refutation, and eventual restriction to prime values of $n + 1$. $n = 12$ provided a further confirming instance. I proceeded to an interactive presentation of a generic proof, inviting Sonia to pick a prime between 11 and 19. She chose 19. I got them to list 1–18 and work on inverse pairs in table-groups, during which Simon spontaneously explained to his colleagues why 18! *had* to be 18 mod 19. I asked him to repeat his reasoning to the class, and wrote his explanation on the whiteboard. He picked out eight inverse pairs and explained why the product of the integers from 2 to 17 inclusive would have to be 1 mod 19. They dutifully copied Simon's argument. Later, I enquired what would have happened if we had looked at 28! mod 29, and Abby explained why it would have to be 28, again referring to inverse pairings of the integers from 2 to 27, although without feeling the need to identify the pairs this time. 'Does everyone agree?', I asked. They agreed. One should not read too much into such consent, however pleasing; nevertheless, Abby, at least, had convinced me that she had appropriated the proof-scheme.

The next day, at a small-group tutorial meeting, I asked five members of the class to write out the proof (that, for primes p, $p! \equiv p\text{-}1 \bmod p$) in conventional generality. Their responses were unaided and individual. It should be borne in mind, as I indicated earlier in this paper, that these students will have had little experience of composing formal proofs. Nevertheless, they all indicated in their writing that the genericity of the case $p = 19$ had been apparent to them. Moreover, their argumentation and use of notation would have satisfied any examiner. Hannah's response, which was typical, was as follows.

$$(p-1)(p-2)(p-3)(p-4)\ldots 2\times 1$$

Every element of M_p[1] has an inverse, because M_p is a group.

We know (from work on primitive roots) that only $p\text{-}1$ has order 2. Therefore, $p\text{-}1$ is self-inverse. All other members of M_p apart from 1 must have a distinct inverse.

Each inverse pair when multiplied gives 1 mod p.

This gives $(p\text{-}1)(1^{1/2(p-3)})1 \equiv (p\text{-}1)! \bmod p$.

Therefore $(p\text{-}1)! \equiv (p\text{-}1) \bmod p$.

Only Zoë gave evidence of some insecurity in this intangible world that lies beyond examples. Her proof was much the same as Hannah's, but began with identification of the inverse pairs in the case $p = 11$, but for Zoë, mere knowledge of the *existence* of distinct inverses in the range 2 to $p\text{-}2$ was not enough. She concluded with the comment: I tried to find a formula for the inverses, for example $p\text{-}2$ has inverse

[1] Mp denotes the group $\{1, 2, 3, \ldots, p\text{-}1\}$ under multiplication mod p.

p-6 (but only for p = 11). I have been unable to do this. What is not clear is whether it leaves *her* cognitively insecure, or whether she believed that I (in my role as assessor) would expect more.

6. Generic arguments and cognitive unity

The domain of elementary number theory lends itself remarkably well to generic argument, presented with the intention of conveying the force and the structure of a conventional generalised argument through the medium of a particular case. One reason for this might be that, in the selection of examples, one seems to be spoiled for choice: there are an awful lot of integers (or primes, or whatever subset is called for) compared, say, with groups, or topological spaces. This is not to say that the choice of a generic example is an arbitrary one: it can be (and in a sense, it *ought* to be) a conscious pedagogical act. Some examples work better than others do for particular purposes—as I indicated in my earlier demonstration of the Euclidean algorithm, a 'good' generic example effectively carries and conveys the generalisation because the salient operations on the variable(s) can easily be tracked through the argument. Some tentative principles for the selection of generic arguments and the construction of generic arguments in number theory are given in Ref. [13]. I conclude, however, with some cautionary remarks.

First, the proof of Wilson's theorem given above crucially depends on knowing and being certain that 1 and p-1 are the only self-inverse elements under multiplication mod p. How shall we establish that result? It emerges readily (as a conjecture, of course) from examples, especially when the contrast is made with non-prime moduli. The usual proof runs as follows: if $1 \leq a \leq p$-1, and $a^2 \equiv 1 \bmod p$, then p divides (a-1)(a + 1) and so p divides (a-1) or p divides (a + 1). Whence a = 1 or a = p-1. The essence of this argument is the solution of a quadratic equation in a multiplicative modular group, which seems to rule out a generic presentation entirely free of algebraic symbolism. It is true that I could take p to be 13 and argue that a must be 1 or 12, but I cannot seem to side-step arguing from (and with) 13 divides (a-1)(a + 1).

Secondly, the converse of Wilson's theorem [if n is composite then (n-1)! $\neq n$-1 mod n] *appears* to lend itself wonderfully well to generic exposition. Take the case n = 10. Now 9! = 362,880, so 9! ≡ 0 mod 10. Yes, but why? Because 9! includes factors 2 and 5. Since 10 is composite it can be decomposed into the product of two factors, both strictly between 1 and 10, so both occur as terms in 9! It is thus apparent that if n is composite then:

$$(n-1)! \equiv 0 \bmod n$$

However, whilst this conclusion is certainly true, the argument does not, in fact, transfer to all composite numbers. In those special cases when n is the square of a prime p, the only possible decomposition of n into the product of two factors, both strictly between 1 and p-1, is $n = p \times p$. The factors are not distinct and it is not the case that both occur as terms in (n-1)! It is not difficult to make a separate argument for these cases, but they can easily be overlooked, and caught in the shadow, as it were, of the earlier, pleasing generic argument.

Notwithstanding these cautionary words, there seems to be a good prospect of developing and offering a systematic didactic technology of formal proof in number theory, building on skilfully constructed generic examples. There is evidence that such an approach to proof is supportive of the *cognitive unity* of theorems, that is to say, 'the continuity ... between the process of statement production and the process of its proof, as well as providing meaningful examples' ([14], p. 182).

7. Conclusion: A challenge to didactical inversion

For all learners of mathematics, there is the possibility of acquiring new knowledge by reflection on appropriate and relevant experience. Unifying and information-extending insight is central to such a means of coming-to-know, and may be viewed as a form of generalisation. MacLane [15] speaks of 'generalisation from cases': a generalisation which subsumes several particular cases, when a general rule summarises some features of the specific cases. For the great mathematicians, as well as for novices, mathematics characteristically comes into being by inductive intuition, not by deduction.

Analysis and natural philosophy owe their most important discoveries to this fruitful means, which is called induction. Newton was indebted to it for his theorems of the binomial and the principle of universal gravity [16].

The purpose of rigour is to legitimate the conquests of the intuition. (Hadamard, quoted in ([17], p.1])).

Burn [11] considered the possibility of the genesis of 'new' mathematics in university mathematics classrooms. He refers to Hans Freudenthal's distinction between three different phenomena in mathematics education:

1. The articulation of formal structures, their logical connections and their applications are the substance of the phenomenon of *mathematics*.

2. The interactions between teacher and student, and the activities engineered by the teacher for the student are a *didactical* phenomenon.

3. The experiences that the student goes through in learning some mathematics are a *genetic* phenomenon of a psychological nature. (p. 21)

Freudenthal writes: 'No mathematical idea has ever been published in the way it was discovered' ([18], p. ix).

Burn comments: 'Freudenthal claimed that mathematical practice often led to a didactical inversion in which the genetic sequence was reversed in exposition. ... The best pedagogy is informed by knowledge of both genetic and formal structure and the difference between them' ([11], p. 21).

It was through the medium of his 'pathway' books – the first of them [15] on Number Theory—that Burn enacted his beliefs about teaching advanced mathematics in the classroom. The chapters of these books consist of a sequence of activities/ problems for students to work on. In a class setting, this could lead to group and/or class discussion, then to some (written) expression of what had been learned from the activity.

When 'new' mathematics comes into being—whether new to an individual or a community, local or international—it marks the end, or a staging point, on a journey. Burn was observing that, when learning more advanced mathematics, students are taken straight to the end of the journey, with little, or no encounter with the path that led to it. If the starting point for teaching is the final outcome of an activity instead of the activity itself, then we see the world upside down. For some 17 years, Burn led a team teaching advanced mathematics in Cambridge, UK, and I was fortunate to be a member of that group. His own approach to mathematics teaching 'rubbed off' on the rest of us. My account of that conversation with Bob Hall about Pythagorean triples captures something of our approach: engage the students in active enquiry, finding examples and generating conjectures. My account of that hour working with first year undergraduates on Wilson's theorem gives more detail of one case of that pedagogy in practice. I have recorded another, concerning the Euler function, in Ref. [19].

Whereas the warrant of a formal proof is logic, the effectiveness of embodied proof by generic example must be determined by the response of the audience for whom it is intended. Thus: logical versus psychological. Exposition versus instruction: it is always beneficial to recognise the difference.

Acknowledgements

This chapter is a development of the author's conference presentation: 'Proofs in Number Theory: history and heresy', given at the 26th Conference of the International Group for the Psychology of Mathematics Education. Norwich. UK: University of East Anglia, July 2002.

Author details

Tim Rowland
Faculty of Education, University of Cambridge, Cambridge, UK

*Address all correspondence to: tr202@cam.ac.uk

IntechOpen

© 2025 The Author(s). Licensee IntechOpen. This chapter is distributed under the terms of the Creative Commons Attribution License (http://creativecommons.org/licenses/by/4.0), which permits unrestricted use, distribution, and reproduction in any medium, provided the original work is properly cited. (cc) BY

References

[1] Tall DO. Using technology to support an embodied approach to learning concepts in mathematics. In: Carvalho LM, Guimarães LC, editors. História e Tecnologia no Ensino da Matemática. Vol. 1. Brasil: Rio de Janeiro; 2003. pp. 1-28

[2] Bruner JS. Towards a Theory of Instruction. Cambridge: Harvard; 1966

[3] Niven I, Zuckerman HS. An Introduction to the Theory of Numbers. 4th ed. New York: John Wiley and Sons; 1980. p. 139

[4] Rowland T, Hall R. The classical form of Pythagorean triples. Mathematical Gazette. 1997;**81**(491):270-272

[5] Reed SK, Dempster A, Ettinger M. Usefulness of analogous solutions for solving algebra word problems. Journal of Experimental Psychology: Learning, Memory, and Cognition. 1985;**11**:106-125

[6] Catrambone R. Following instructions: Effects of principles and examples. Journal of Experimental Psychology: Applied. 1995;**1**(3): 227-244

[7] Mason J, Pimm D. Generic examples: Seeing the general in the particular. Educational Studies in Mathematics. 1984;**15**:277-289

[8] Balacheff N. Aspects of proof in pupils' practice of school mathematics. In: Pimm D, editor. Mathematics, Teachers and Children. London: Hodder and Stoughton; 1988. pp. 216-235

[9] Polya G. Mathematical Discovery. Vol. I. New York: John Wiley; 1962

[10] Hoffman P. The Man Who Loved Only Numbers. London: Fourth Estate; 1998

[11] Burn RP. The genesis of mathematical structures. In: Kahn P, Kyle J, editors. Effective Learning and Teaching in Mathematics and its Applications. London: Kogan Page; 2002

[12] Rowland T. Proofs in number theory: History and heresy. In: Cockburn AD, Nardi E, editors. Proceedings of the 26th Conference of the International Group for the Psychology of Mathemati0cs Education. Vol. 1. Norwich, UK: University of East Anglia; 2002. pp. 230-235

[13] Mariotti MA, Bartolini Bussi M, Boero P, Ferri F, Garuti R. Approaching geometry theorems in contexts. In: Pehkonen E, editor. Proceedings of the 21st Conference of the International Group for the Psychology of Mathematics Education Volume 1. Lahti, Finland: University of Helsinki; 1997. pp. 180-195

[14] Rowland T. Generic proofs in number theory. In: Campbell S, Zazkis R, editors. Learning and Teaching Number Theory: Research in Cognition and Instruction. Westport, CT: Ablex Publishing; 2002. pp. 157-184

[15] MacLane S. Mathematics: Form and Function. New York: Springer-Verlag; 1986

[16] Laplace PS. A Philosophical Essay on Probabilities. New York: Truscott and Emory; 1901

[17] Burn RP. A Pathway into Number Theory. Cambridge: Cambridge University Press; 1982

DOI: http://dx.doi.org/10.5772/intechopen.1012302

[18] Freudenthal H. Didactical Phenomenology of Mathematical Structures. Dordrecht: Reidel; 1983

[19] Rowland T. Mathematics teaching - joy in times of trouble? For the Learning of Mathematics. 2021;**41**(3):25-29